Your Wind Driven Generator

Your Wind Driven Generator

Benjamin Wolff
with
E. James Brabant

for
North Wind Power Company, Inc.

VNR VAN NOSTRAND REINHOLD COMPANY
NEW YORK   CINCINNATI   TORONTO   LONDON   MELBOURNE

Copyright © 1984 by Van Nostrand Reinhold Company Inc.

Library of Congress Catalog Card Number: 83-5772
ISBN: 0-442-29336-4
ISBN: 0-442-29337-2 pbk.

All rights reserved. No part of this work covered by the copyright hereon may be reproduced or used in any form or by any means-graphic, electronic, or mechanical, including photocopying, recording, taping, or information storage and retrieval systems-without permission of the publisher.

Manufactured in the United States of America

Published by Van Nostrand Reinhold Company Inc.
135 West 50th Street
New York, N.Y. 10020

Van Nostrand Reinhold Company Limited
Molly Millars Lane
Wokingham, Berkshire RG11 2PY, England

Van Nostrand Reinhold
480 Latrobe Street
Melbourne, Victoria 3000, Australia

Macmillan of Canada
Division of Gage Publishing Limited
164 Commander Boulevard
Agincourt, Ontario M1S 3C7, Canada

15 14 13 12 11 10 9 8 7 6 5 4 3 2 1

Library of Congress Cataloging in Publication Data

Wolff, Benjamin.
  Your wind driven generator.

  Bibliography: p.
  Includes index.
  1. Wind power. I. Brabant, E. James. II. North Wind power Company. III. Title.
TK1541.W64 1984   621.31'2136'0687   83-5772
ISBN 0-442-29336-4
ISBN 0-442-29337-2 (pbk.)

FOREWORD

Wind, the movement of our invisible atmosphere, has fascinated man for ages. Mild one moment, strong the next, sometimes imperceptible, sometimes tempestuous, but constantly surrounding us. The fascination with the wind is deeply rooted. Every culture has names for the wind in its different moods; Scirocco, Chinook, Foehn, Noreaster, Typhoon, Zephyr. We name our airplanes and even our automobiles after them.

But man is as much a pragamatist as a romanticist. He needs power and energy as much as he needs tools and materials to control and maintain his life. The power in the wind once experienced, is easily considered, but not easily tamed as a source of energy. With the possible exceptions of muscle-power and burning wood, wind is however the oldest form of energy to be captured and used by man. Up and down rivers and across the seas, the sailing ship helped build ancient civilizations.

An inventor, his name lost to history, smarter or more observant than his peers, created the first windmills almost 20 centuries ago in Persia. A mud-brick shelter with slots for the wind to enter and rotate a vertical paddle wheel could grind grain without tiring. This wind-driven machine evolved over 2,000 years. In the seventh or eighth century, the propeller type windmill was invented. In the eleventh or twelfth, the Crusaders brought the idea to western Europe. For several centuries, the windmill became the mainstay of stationary power in northwestern Europe, and early settlers brought the technology to the New World.

Putting the wind to work, that ancient inventor must have asked himself, "Where does the energy come from?" Our modern answer is that it originates with the sun's uneven heating of the atmosphere due to latitude and varying surface reflectivity and absorbtion characteristics, as modified by the rotation of the earth and local topographical featues. An equally valid answer is that "it comes from a machine." That machine takes the energy in the form Nature provides and converts it into a form useful to meet human needs. As a machine, it must not only follow the "laws" of physics, but all the "laws" of man—institutional, environmental, and economic. Particularly economic, for if an energy source is not economic, it will soon be replaced by another source. The history of wind power, whether for transportation or stationary power, shows its use to have been cyclical. It has been widely used or replaced, depending on the relative economics between it and its competition—coal, oil, peat, or galley slaves.

When the familiar "Dutch windmill" technology was brought to this country, and there are still existing examples, it failed economically. In the southeast, the winds were too low. In the northeast, the winds were strong, but the "New England" had many fast running streams for water wheels, and the cloth covered sails were too labor intensive for a population-short colony. It wasn't until the 1800s, when the multi-bladed water pumping windmill, with its automatic mechanism to regulate speed without a human operator was invented, that wind power succeeded in this country, changing the color of the great plains from brown to green, hastening the settling of the West. The use of wind decreased with the advent of coal- and oil-fueled steam engines, but came back in a new form—the propeller-type, wind-driven generator—in the 1920s to provide the first electricity for many of our nation's farms. In the 1930s the Rural Electrification Administration, bringing utility power to the dispersed farms, made wind uneconomical yet again.

In the early 1970s a few people, a very few, became interested in wind power. In times of change there are always a few key individuals that through foresight, insight, and imagination identify a problem and have the spirit and courage to attack it. In those early days of modern wind development, it was a few analysts, a couple of university professors, and a small group of young entrepreneurs. Ben Wolff was one of them. They were the ones willing to stake their time, skills, careers and personal resources to hasten the next cycle of the practical use of wind power. Out of a commitment to energy and environmental concerns at a time when OPEC sounded like a new grocery chain to most people, they began trying to accelerate what usually would be a fifty-year cycle down to ten years. Small companies were formed, some rebuilding machines from the 1930's, some importing French, Australian, or Swiss machines (there were only three manufacturers in the world at that time), and some trying to develop new designs.

At that time, I was on loan from Cal-Tech to the National Science Foundation, in career transit from engineer to bureaucrat. A few policy studies had already shown that energy was going to be a long-term problem. Research, on a small scale, was started on renewable energy sources, and wind was being studied as a subset of solar energy.

I remember my first meeting with Ben and other members of Windworks. It was a culture clash not much less than that between Cortez and the Aztecs. Young enthusiastic engineers meet government high technology and research laboratories. Then the 1974 oil embargo hit. We learned to learn from each other. Many a discussion of problems and needs of an embryonic industry would lead to a glimmer of an idea, a start toward a solution to the problem of the moment. Out of that experience and over the last ten years has grown a mutual respect that comes from long hours pursuing a difficult, mutual goal.

An enormous change has occurred in the field of wind energy over the past ten years. What was considered by many as one of the less likely candidates for a viable energy source has turned into a rapidly growing, economic in the right location and application, and practical source of energy. Over 30 manufacturers are now in business developing and selling small wind machines for homes, farms, or other applications. A dealer-distributor network is forming around the country. The fundamental problems of interconnection with the utility system have been solved, and small wind systems are beginning to be used in nearly every state and territory.

This is not to say there are no problems involved in buying and using a wind machine. Quite the contrary. It is a machine and, while not as complex as an automobile, still requires understanding and respect. A fifteen-foot blade turning at 150 rpm on top a 60-foot tower and attached to a 220-volt electrical line is no toy. There are excellent machines available but at the same time, with no industry-wide standards and "brand names" not yet evolved, there are some that, to be charitable, "come and go"—as do some manufacturers.

Distinguishing between various machines and manufacturers, knowing what questions to ask, becoming a knowledgeable user, or potential user, is what this book is all about. Ben Wolff is eminently qualified to present that information. He has built and operated systems, from designing to climbing the tower. As Executive Director of the American Wind Energy Association, he helped resolve the myriad of institutional and political problems involved when wind systems first began to be used by the public and connected to a utility.

In many fields of endeavor there is a gap in the literature; highly technical on one hand, too simple to be of real use on the other, and precious little in the middle. I have some 18 linear feet of technical reports on wind energy in my office, vital for designing and advancing the technology, but most totally unuseable by any but the specialist—there are some I don't even understand. There are also a number of popular articles and books, but most too elementary to be of serious use.

This book fills that gap and does it in a clear and lucid style; technical enough to be useful and simple enough to be practical. It provides the information necessary to decide whether and how to buy and use a wind machine and includes not only technical factors, but institutional factors as well. It does so objectively, pointing out the pitfalls, risks and problems as well as the advantages and rewards, and does so from a large base of experience. This book is a must for anyone interested in truly understanding and harnessing one of the most simple, yet complex forms of renewable energy.

The past ten years have seen a tremendous change in the technology, the understanding, and the view of wind power; it has been an exciting decade. I believe the next decade will be equally exciting in the commercial development and practical application of wind power, and this book should be one of the charts through it.

Louis V. Divone
Vienna, VA

ACKNOWLEDGMENTS

The original impetus for this book was provided by North Wind Power Company, who, in an effort to respond to the flood of requests for information on residential wind energy systems, commissioned this book. Particular thanks are due to Donald Mayer, John Norton, and Jito Coleman for the many hours spent reviewing the many drafts.

Additional thanks are due to E. James Brabant and Thomas V. Vonier whose efforts at rewriting and illustrating contributed greatly to making the text comprehensible.

Many other reviewers of early drafts helped clarify and concentrate the author's attention including Wiley Wilson, Peter Smealie, Rick Katzenberg, and Tony Clifford.

Thanks are also due to Louis Divone for contributing the Foreword, and Department of Energy sponsored researchers whose work contributes to important parts of this book.

While the mistakes are solely the author's, that the book is readable at all is due in large part to the tireless work of my editors at Van Nostrand Reinhold, Susan Munger and James Costello, and the extensive assistance of Karl Ackerman.

Finally, this book would not have come into being without the loving support and patience of my wife Anne.

CONTENTS

Foreword	iii
Introduction	ix
1. Energy: The Basics	1
2. The Wind as a Source of Energy	8
3. Is Wind Energy Practical?	19
4. Choosing the Right Wind Machine	35
5. A Wind Machine Investment	50
6. The Small Wind Systems and the Utility	60
7. Owning a Wind Machine	74
8. One Family's Experience	87
9. Postscript	91
Selected References	97
Glossary	99
Appendices:	
Available State Incentives	107
Present Value Tables	159
Technical Wind Energy Bibliography	165
Commercial Wind Machines	177
Index	181

INTRODUCTION

In which the author describes his experience in the field of wind energy, which parallels the recent development of wind energy technology, and outlines the purpose of this book.

Many years ago, R. Buckminster Fuller, architect and inventor, realized that the atmosphere is the largest, most efficient storage system of solar energy available to mankind. The world-famous designer of the geodesic dome also knew it took at least ten years for his ideas to be put to practice. Still, he decided to provide a small monthly grant to a group of engineers, architects and craftsmen to develop the technology to tap the immense potential of the wind. This group, of which I was a member, eventually became known as Windworks.

Hans Meyer, a long-time associate of Bucky's, had begun investigating and applying new materials and the latest research in aerodynamics to the problem of extracting usable energy from the wind. These early experiments produced a prototype system using a shroud around the blades to increase the power output, named a *Venturi* after its inventor, and the use of paper honeycomb to make wind machine blades (Figures 1 and 2).

An early goal was to design wind machines that could be built with simple tools in a home workshop. One design was published in November 1972 in *Popular Science* with the wind machine featured on the cover. At the time, it was the best selling issue in the magazine's long history, and the editors attributed it to the public interest in wind energy. In the months following the article our mail jumped to fifty letters a day—each one asking for more information on wind energy.

There were many successes and failures in those early years of home-built wind machines (Figure 3). The excitement of carefully fabricating the pieces, assembling them and erecting the machine and finally watching the ammeter and voltmeter as electricity was generated was tempered with drive trains that simply didn't work, fractured blades and even fallen towers.

After building five or six prototypes, we became convinced that the design and construction of a durable, high performance wind system was an extremely complex task, beyond the capability of the home craftsman.

By this time, people didn't laugh when they learned we were working on wind energy, and many other small businesses exploring this revitalized technology sprang up around the country. North Wind Power located in Vermont was one of these companies. North Wind studied the wind driven generators common in the United States before the advent of the Rural Electrification Administration (REA). Before the REA brought power lines to the rural areas of the United States many farmers generated electricity from small

Fig. 1. Venturi.

Fig. 2. Honeycomb blades.

Fig. 3. Early prototype.

wind machines for radios or a few lights. These machines, from a few watts to two or three kilowatts, produced direct current electricity which was stored in batteries for periods of no wind. North Wind visited North and South Dakota, Minnesota, Wyoming, and Montana and collected hundreds of machines from more than ten manufacturers. Many of these machines had been abandoned on their towers, or heaped on the piles of rusting equipment found around the farm. A rich lore surrounded these old wind machines and was shared by the farmers around the wood stove.

"Yep. Canadians came through here a few years ago and picked up all the 110 volt machines."

"There was this radio, Zenith I think it was, when I got the radio, they gave me a wind generator to power it for free."

"The only problem I ever had with mine was the slip rings (used to bring the electricity to the bottom of the tower while allowing the machine to rotate 360° to respond to changes in wind direction) I'd say they needed replacing every coupla years."

Fig. 4. Drive train and alternator.

"I still don't know what the wind speed is around here. My neighbor got one and it worked for him, so I got one."

"A lot of wind companies went out of business during the war. Manufacturers were being rationed steel, and these companies got bought out in order to get their materials ration."

In 1975 the group I was working with became incorporated as Windworks, Inc. Windworks, also made a number of trips west to collect old wind machines. We rebuilt, erected and tested the performance and reliability of these old machines. Though the technology appeared simple, it was obvious that wind generators were sophisticated machines that required careful design and manufacture. There seemed to be a limit in the size of machines built in this period, approximately 16 feet in blade diameter and two or three kilowatts in capacity. Most were smaller. We concluded it was probably because the structural loads on the machines, particularly the blades, were poorly understood. We heard of one designer who would climb the tower to stand behind his machine when the wind was blowing. He would then try and light a match. If he was successful, he figured he had a good blade design because it was stopping so much wind. Today, blade design is much more sophisticated, involving many hours of computer analysis.

The greatest variation in machine designs of the 1930s occurred in the approaches that were taken to prevent damage to the machine in high winds (Figures 5, 6, and 7). These *overspeed controls* included machines with paddles used as "air brakes," machines with weights attached to the blades, and others that would "flip up" when the winds got too strong. By rebuilding, testing, and installing these machines we learned what did and didn't work. We quickly learned the importance of quality control, making certain suppliers provided bearings, castings, and other parts that were correct. We experienced many of the problems associated with installing machines and keeping them operating.

Over time, it became clear that the most efficient and reliable designs had the fewest number of moving parts, for example, direct drive generators with the rotor linked directly to the generator rather than using mechanical power transmission. The control systems that prevent damage to the wind machine in high winds were most effective when they were simple and direct.

During the early 1970s we realized that wind energy would not be competitive with conventional utility electricity unless we could eliminate the costly and expensive batteries that stored electricity for periods without wind (Figure 8).

The question most often asked about wind energy was "What do you do when the wind doesn't blow?" Battery banks provided an adequate technical solution, but not an economical one. On a bright sunny day, a

Fig. 5. Fixed blade with air brakes: 1920s.

Fig. 6. "Fly-ball" hub: 1930s.

Fig. 7. Variable axis rotor: 1930s.

Fig. 8. Typical battery bank.

station wagon pulled into the driveway at Windworks. Alan Wilkerson, an electronics expert and inventor, got out with a maze of solid state devices, a printed circuit board, and large heat sinks hastily wired together and mounted on a piece of plywood. Al had read an article in Smithsonian Magazine about our work on wind energy and came out for a visit. The device he brought with him enabled the electrical current from the wind machine to be fed directly into the utility grid. With this device, when the wind machine was not producing enough energy to meet the needs of the user, the difference was automatically drawn from the utility. On the other hand, when the wind machine produced more power than was needed, the excess could be fed into the utility, and the electric meter would run backwards. With the battery storage system eliminated, the cost of wind-generated electricity was suddenly and dramatically cut in half.

Representative Henry Reuss (Democrat—Wisconsin), then chairman of the House Banking Committee and known for his foresightedness and his sensitivity to environmental concerns, read the same article. He called us to see if a wind machine would be practical for his property in Wisconsin. In less than a year, Representative Reuss had the first commercial wind machine capable of selling power to the utility (Figure 9). Speaking before various groups, he would gleefully rub his hands together and announce, "I'm waiting for the day when the electric company reads the meter, comes to the door, and says, 'Mr. Reuss, here's your check for $3.75'."

Since that time, many other devices and approaches have been developed to connect an individual wind machine to the utility grid. Today, a small wind machine operating in parallel to the utility grid is the most

Fig. 9. Congressman Reuss's wind machine.

widespread application of wind energy because of refinements in interconnection hardware and new understanding of the impacts of wind machines on the utility's operation.

Until the oil embargo, the government paid little attention to wind energy. In one instant, the nation became aware that our conventional energy supplies were finite, and more important, that our domestic demand for energy had outstripped our domestic supply. The government began an accelerated research and development program in a number of unconventional energy technologies: solar heating and cooling, photovoltaics, ocean thermal energy conversion, and wind energy. The Department of Energy sponsored a "design competition" to encourage industry to design machines in various sizes for a range of applications. All the wind companies and any other interested firms were invited to submit proposals for new designs that would employ the latest advances in aeronautical, mechanical, structural and electrical engineering. Wind machines developed under these contracts ranged from 1 kilowatt to 40 kilowatts in capacity. Of the hundreds of proposals that were submitted, only a handful of companies were awarded contracts, including Windworks and North Wind (Figure 10).

Most of these designs have been built and are being tested. Although each design employed extensive computer-assisted analysis, advanced structural engineering, and the latest research in aerodynamics virtually all the early prototypes experience a range of problems, from broken parts to performance substantially below predicted levels. Some designs have been scrapped, some significantly modified. Continued testing and operating experience with these designs and others developed without government support are establishing the basis of continued improvements, reliability and performance in the modern wind machine.

In 1978, I moved to Washington, DC to establish a national office for the American Wind Energy Association (AWEA). Initially, two activities consumed a tremendous amount of time; providing accurate, reli-

APPLICATIONS	ANNUAL ENERGY REQUIREMENTS (kWh/yr)	SIZE (kW)
• REMOTE NAVIGATION AIDS GALVONIC PROTECTION, HIGHWAY LIGHTING	2,000–7,000	1–2 (HIGH RELIABILITY)
• RESIDENTIAL SINGLE-FAMILY RESIDENCE, (EXCLUDING SPACE HEATING)	7,500–15,000	4
• RESIDENTIAL/FARM SMALL FARM, RESIDENCES (INCLUDING SPACE HEATING)	30,000	8
• FARM/COMMERCIAL POULTRY, LIVESTOCK LIGHT IRRIGATION, COMMERCIAL ENTERPRISE	45,000–55,000	15
• FARM/COMMERCIAL DEEP WELL IRRIGATION, ISOLATED COMMUNITIES, COMMERCIAL OPERATION	120,000	40

Fig. 10. The "Design Competition" wind machines.

able information on wind energy to the legislators on Capitol Hill and to the growing number of companies and individuals who suddenly found the prospect of using wind energy very attractive.

There were long gas lines around the country, and serious concerns about domestic energy supplies, but there was a growing awareness that even the best-intentioned government program might do more to hurt a fledgling industry than help it. One official at the Department of Energy noted: "The government is like a friendly elephant who wants to help, but you have to be careful that it doesn't roll over on you."

The near total lack of information on wind energy in the nation's capital required looking closely at wind energy in the context of other energy technologies, as well as at the broad role of the federal government. The formulation of national energy policy and the role of wind energy within that policy has social, economic, and even national security implications. When a representative of the Department of Energy, during testimony before the Senate Energy Committee, told the senators that wind energy was too expensive to provide much of a contribution before the year 2000, one senator testily responded; "Could you tell me sir, how much it will cost to send the Marines to the Middle East?"

These and other issues were considered and debated; and compromises were sometimes reached. At other

Fig. 11. A wind farm.

times, the political power of entrenched energy interests was felt, and all the logical arguments about wind energy and other alternatives could have only minimal impact on the political realities of energy politics. In today's political environment, energy policy will be hammered out in the marketplace—energy consumers will vote with their wallet. Recent advances in alternative energy technologies, not all of them government-funded, have ensured a choice.

While energy policy was being debated in Washington, many individuals started looking for alternatives to their climbing utility bills. Many had been caught in the utility's version of Catch-22; you use less you get charged more. The American Wind Energy Association was the starting point for many investigating the possibilities of using wind energy. Calls and letters requesting information reached 300 per day. While many basic questions were asked over and over, there was a change. The public was becoming more knowledgeable about energy in general and wind energy in specific. The questions changed from "What do I do when the wind doesn't blow" to "How does marginal cost pricing affect the rate utilities will pay me for electricity from my wind machine?" Many individuals had read magazine or newspaper articles on wind energy, received pamphlets from manufacturers, or reports on government-sponsored research. Too often, they found that either they still didn't have enough information, or too much of the wrong information.

This book is an attempt to provide the future users of wind energy with answers to their questions, or in some cases, the correct questions. The focus is on the most common questions about energy production, utilities, zoning, and economics. Some of the information is technical in nature and demands careful study by the reader. It's worth it. The effort expended can help avoid expensive mistakes and ensure a successful wind machine installation (Figure 11).

The momentum of the wind industry is slowly, but inexorably building. The role of the government, invaluable in the past, is unlikely to be critical in the future. There is little the government can do to make people use wind energy or protect them from bad investments. It is risky to try to predict the future, but one thing is certain. Informed citizens will play a pivotal role in shaping our energy future. Individually we will make choices that will shape our energy future. As Isaac Asimov, the noted science writer said: "However pleasurable spectatorship may be, participation is better."

Your Wind Driven Generator

1
ENERGY: BASIC FACTS

The attraction of alternate energy; the characteristics of conventional and renewable energy sources; the uses of energy in the United States and the home; the role of wind energy and the current state of wind energy technology.

The energy crisis seems to have vanished as suddenly as it appeared. The gas lines, mandatory temperature levels in public buildings, energy performance standards for new buildings—they're all gone. And so is our energy innocence. We have learned that electricity doesn't come out of a hole in the wall and that our oil supply lines are all too fragile.

The crisis remains, if not on the front pages of the newspaper, in escalating oil, gas, and electric bills. As energy costs become a larger portion of the family or business budget, conservation begins: insulation is installed, storm windows go up, unnecessary lights and appliances are turned off and thermostats are turned down. This is the most economical and immediate action that can be taken to reduce energy bills. However, even with reduced use, utility bills continue to climb. The rising cost of energy is the largest single factor in the growing interest in the use of wind energy (Figure 1-1).

The availability of energy has become a serious concern. Shortages of natural gas, oil, and coal have already caused layoffs and school closings in some parts of the country, and brownouts threaten in other parts. Even the current terms used to describe adequate supplies—gas *bubbles* and oil *gluts*—imply future shortages. Americans are dependent on foreign oil and volatile politics of the Middle East. Conspiracy theories are advanced as an explanation for the apparent lack of accountability in the energy marketplace. Those not attracted to conspiracy theories fear that perhaps no one is in control (Figure 1-2).

Cost, availability, and dependence force many to explore the potential of alternative sources of energy which can be controlled by the user. Solar, wind, and biomass are only a few of the options available.

There are many different sources of energy and each has unique characteristics. For example, electricity is not widely used to power automobiles, and nuclear power doesn't directly heat homes or offices. The most widely used types of energy today are fossil fuels: coal, oil and natural gas. They share some specific characteristics:

- *They are available on demand.* The firing of a spark plug releases solar energy stored a millenium ago. When winter arrives, a telephone call fills the oil heating tank.

Fig. 1-1. Eat or heat.

- *They are dense.* In the space of our gasoline tank, enough energy is stored to propel a two-ton automobile over 300 miles.
- *They are finite.* The energy we release from fossil fuels was stored in the era of dinosaurs, and new fossil fuels are not being created. Experts estimate that we have already used more oil than is left under the ground.
- *They pollute.* Some pollution is direct: from burning hydrocarbons, or from incomplete combustion. Some pollution is less direct: lake-killing acid rain, the result of pollution from Midwest power plants, or accumulating carbon dioxides which are slowly raising the temperature of the earth and which the scientific community increasingly believes may lead to the melting of the polar icecaps if not reversed.

Sources of energy that are commonly referred to as alternative sources include solar, wind, biomass, and hydropower. They share virtually the opposite characteristics of conventional fossil fuels:

- *They are intermittent.* Although we can be confident that the wind or sun will be available for a certain portion of the year, they may not be available precisely when we want to turn on the toaster.
- *They are dilute.* One barrel of oil, only 42 gallons, contains the energy equivalent of approximately 1,699,000 kilowatt-hours of electricity. Solar energy falling on one square meter of land at sea level is

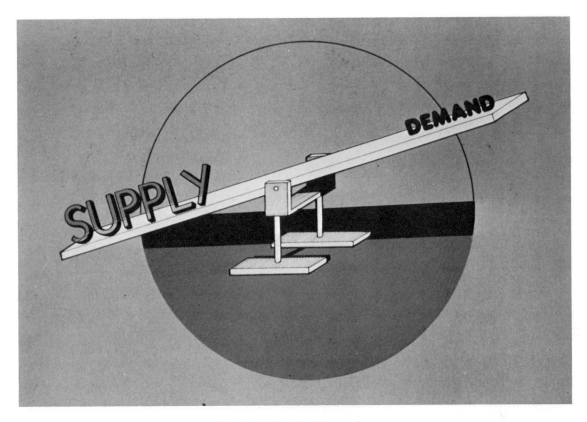

Fig. 1-2. Tipping the scales.

approximately 1 kilowatt. It would take 195 square meters (2,087 square feet) of solar energy for one year to equal the amount of energy in a single barrel of oil.
- *They are inexhaustible.* Direct and indirect solar energy will be available as long as the sun shines—several billion years.
- *They do not pollute.* Because there is no direct combustion involved in the use of most alternative sources of energy, there is no direct pollution. No nitrous oxides, oxides of carbon or sulfur. There is virtually no indirect pollution, no strip mines or waste.

These are the most general characteristics of conventional and alternative sources of energy. Each type has additional specific characteristics. For example, solar collectors may work well to heat space or water, but not to power a table saw. Both the general and specific characteristics of various forms of energy should be considered in the context of how we use energy.

In 1978 the United States consumed 78 quads of energy. A quad equals a quadrillion British thermal units, or 10^{15} Btu. One quad is:

- Enough energy to heat 500,000 homes for twenty years (based on an 1800 ft. house in the Washington, D.C. area).

4 YOUR WIND DRIVEN GENERATOR

- Enough crude oil to fill a fleet of seventy-five supertankers (based on 325,000 ton tankers, each with a capacity of 2.3 million barrels of oil).
- A mountain of coal, 1,000 ft. high and 2,000 ft. in diameter; or 500,000 railroad cars full of coal (based on 83 tons per car).

A breakdown of the energy consumption in quads by sector of the economy and by fuel type is shown on Table 1-1.

Most modern wind machines are designed to produce electricity. Although 30 percent of the energy used in the United States is converted into electricity, only one-third of that performs useful work. The rest is lost in generation or transmission. For example, there are 22 million Btu in a ton of anthracite coal. When it is burned to produce steam to drive turbines and turn generators to produce electricity, energy is lost in each stage. Heat is lost when water is converted to steam and steam escapes instead of turning the blades of the turbine. Mechanical energy is lost while transmitting power to the generator. And electricity is lost in the transmission wires. Even though we started with 22 million Btu in the ton of coal, the energy equivalent of only 7.26 million Btu is used in homes or businesses. For every kilowatt-hour of energy generated in a wind machine, three kilowatt-hours of conventional fuel is saved.

Electricity is a "high-grade" form of energy. It can be used for many different tasks, from lighting the streets to providing mechanical power of motors. It is clean at the point of consumption, and is easily transported from place to place. No other form of energy shares all these characteristics. Many conventional forms of energy—coal, oil, and natural gas, are often converted into electricity. As a result, electricity is the most expensive form of energy used in most parts of the country.

Residential and commercial uses consume over half of all the electricity generated in the United States. In homes electricity is used for lights and a few appliances and maybe for space heating, air conditioning, and

Table 1-1. Energy Consumption (quads) by Sector and Fuel Type—1978.

	COAL	NATURAL GAS	OIL	HYDRO	COKE IMPORT	ELECTRICITY DISTRIBUTED[1]	ELECTRICAL ENERGY LOST[2]	NUCLEAR	OTHER	TOTAL
Residential and Commercial	0.265	7.678	7.227			4.079	10.098			29.347
Industrial	3.433	8.284	6.814	0.036	0.131	2.731	6.769			28.198
Transportation		0.538	20.017			0.015	0.037			20.607
	3.698	16.500	34.058	0.036	0.131	6.825	16.904			78.152
Utilities	10.372	3.297	3.906	3.109				2.977	0.068	23.729

1. Electricity Distributed refers to delivery energy
2. Electrical Energy Lost refers to energy lost from original fuel to end use

- Residential & Commercial: Energy consumed by private households primarily for space heating, air conditioning, and cooking. Also, energy consumed by non-manufacturing businesses including, motels, restaurants, wholesale businesses, retail stores, laundries as well as health, social, and educational institutions and energy consumed by federal, state and local government.

- Industrial: Energy consumed in manufacturing, construction, mining, and farming.

- Transportation: Energy consumed to move people and commodities in both the public and private sector, including, military, railroad, and marine uses, as well as the pipeline transmission of natural gas.

- Utilities: Energy consumed by privately and publicly owned establishments which generate electricity primarily for resale. Note that "Utilities Total" equals the sum of "Electricity Distributed" and "Electrical Energy Lost".

Table 1-2. Electricity Consumption—Household Appliances.

APPLIANCES	CONSUMPTION (kWh/yr)
Refrigerator—auto defrost	1,500–1,600
—manual	950
Range	670–1,175
Freezer	1,380
Clothes dryer	990–1,050
Lighting	700–1,100
Color TV	290–400
Miscellaneous	1,000–2,000
Water heating	3,725–6,400
Air conditioning—room	500–2,000
—central	1,300–3,560
Space heating —resistance	7,000–30,000
—heat pump	8,150–15,000

hot water. The following tables show both the annual consumption of a number of common household appliances and applications. The range is extremely broad from 4,000 kWh/yr for lights and appliances to 56,000 kWh/yr for space heating, hot water, lighting and appliances (Tables 1-2 and 1-3).

To assess the feasibility of using wind energy for your home or business, you must know exactly the amount of electricity you use. Detailed lists of the energy used by various appliances are available, and your local utility can be helpful in assessing your electricity requirements.

THE ROLE OF WIND ENERGY

The limits and characteristics of wind energy should be understood to avoid unrealistic expectations. It is intermittent, essentially not available on demand, and very dilute. However, it is constantly renewed by incoming solar energy, and is available for all practical purposes, forever.

In 1976, at the beginning of the federal wind energy program, the government sponsored a study to determine the potential of wind energy. General Electric, which conducted the study, estimated the wind could contribute over 13 quads to national energy supplies—more than 20 percent of current needs. Other estimates run as high as 30 quads. General Electric found the electric utility market was the largest in potential wind energy production. They projected that 313,500 machines of 1,500 kWh capacity could be deployed to generate 1,070 billion kWh annually, at a savings of 10.7 quads of conventional fuels.

Table 1-3. Electricity Consumption—Typical Applications.

APPLICATION	TYPICAL ANNUAL CONSUMPTION (kWh/yr)	TYPICAL CONSUMPTION RANGE (kWh/yr)
Lights and appliances only	7,150	4,000–8,000
Hot water, lighting and appliances	15,100	9,000–16,000
Space heating, hot water, lighting and appliances	41,600	19,000–56,000
Space heating and hot water only	34,500	15,000–48,000

A study in 1981 examined the residential market for small wind machines, found over 17 million potential residential applications for single family residences in the United States if the machine cost was reduced to a level consistent with mass production.

These numbers are abstract, but they have important implications for an individual considering the use of wind energy. If the residential use of wind energy is only a fraction of the estimates, the production and distribution of machines will become a major industry. However, in 1982 there were less than 3000 wind machines installed at residences around the country.

This situation is similar to the early days of the automobile when a large number of manufacturers supplied a small number of machines. The Pierce-Arrow was essentially hand-crafted and relatively expensive. Until Henry Ford introduced the techniques of mass production, cars were not widely affordable. Local authorities were unfamiliar with the new technology of the horseless carriage. In the early days, some towns had laws requiring that an automobile be proceeded by a man carrying a red flag because of the danger of explosive gasoline. Farmers unable to get their machines repaired, made their own parts casting their own bearings, and liberally applying "spit and bailing wire." Until better roads were built, the car was more of a novelty than a form of transportation.

Manufacturers will have to produce affordable and reliable wind machines. A few pioneers will have to be the first in their area to explain wind energy to local zoning boards, building inspectors, and their neighbors. The purchase rate of power by a utility will have to be negotiated a hundred times before it becomes commonplace.

In the 1930s wind machines on remote farms had little competition. Now they compete with convenient, relatively economical electric utility power. While electric rates are increasing, electric bills remain a small part of a household's monthly outlays. When faced with a decision to continue to pay a monthly electric bill of $20 to $150 a month or spend $5,000 to $30,000 for a wind system, the electric bill option is usually selected.

Wind machines are economical only if specific conditions are met. The durability of new machines is unproven and potential users are not familiar with having their own power plant in the back yard.

There have been a variety of problems over the last few years—years of embryonic growth for the wind industry. Some distributors have collected deposits on wind machines, and then gone bankrupt. A machine may break, and it may take months to get replacement parts from the manufacturer. Unreliable machines have been sold to eager, but unsuspecting customers. Lightning has struck a utility line, flowed through the transmission lines, and blown out a wind machine's electronics.

Proper installation has been a problem for wind machines as well as solar collectors. There is a story of a Hollywood actor who purchased a solar hot water heater from a firm in Australia. Because Australia is in the southern hemisphere, the installation instructions said that the collector should be pointed north—which the actor did. While it is difficult to install a wind machine in anything less than an upright position, it is not uncommon for a machine to be installed in a area where there is virtually no wind–and virtually no power generated. One wind machine owner, writing in the New York Times, admitted to feeling unsettled when the installation crew celebrated the erection of her machine. Did they know something that she didn't?

Despite early problems, more than three thousand modern wind generators are installed in the United States, most spinning with each breeze, delivering clean inexhaustible electricity to their owners (Figure 1-3). The wind energy industry is in a period of phenomenal growth, spawning many new products and

Fig. 1-3. Residential and utility-class wind machines.

companies. Everyone is learning more, and mistakes are less frequent. Though the principles of tapping the energy in the wind won't change, using renewable sources of energy will place new demands of understanding and involvement on the potential user. And with a wind machine that costs thousands of dollars, only an informed consumer will become a competent wind energy producer.

2
THE WIND AS A SOURCE OF ENERGY

Wind energy characteristics; temporal and spatial variations, energy density, environmental impacts—and their implications for the potential wind machine user.

"What do you do when the wind doesn't blow?" though a simple question, implies an understanding of one of the primary characteristics of the wind—its variability. Although no switch turns on the winds, they are remarkably constant over long periods of time and great distances. But the wind is extremely unpredictable over short periods of time and small distances. These differences in time and space mean the wind changes both speed and direction in fractions of a second. The change of a few feet can mean the difference between strong wind and no wind.

We can be confident over a year's time that a certain amount of wind will be available at a given site, but the variations from month to month are likely to be substantial. In many parts of the country, there may be no wind in July and August, but plenty during the changing seasons of spring and fall. There can also be dramatic changes from day to day and hour to hour.

Daily and hourly changes have many causes. Major weather fronts moving through an area may result in strong winds. Large bodies of water can result in differences in air temperature and therefore air pressure over the land and the water causing sea breezes (Figures 2-1 and 2-2). Valleys and mountains heat up and cool off at different rates creating predictable wind patterns (Figures 2-3 and 2-4). An example of these

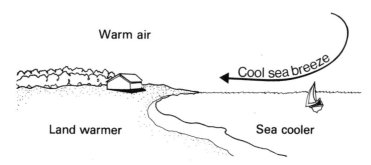

Fig. 2-1. Land/sea breeze—daytime.

THE WIND AS A SOURCE OF ENERGY 9

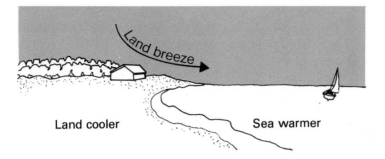

Fig. 2-2. Land/sea breeze—nighttime.

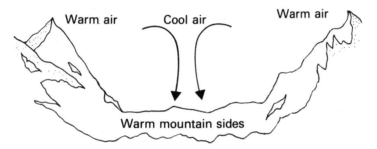

Fig. 2-3. Mountain/valley breeze—daytime.

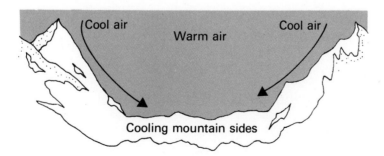

Fig. 2-4. Mountain/Valley breeze—nighttime.

local terrain features are mountain passes in California. During the day, the desert air heats up and rises, leaving a partial vacuum that pulls the cooler air over the ocean through the mountain passes. These winds are predictable and at times very strong and many wind farms are being constructed in California to take advantage of this terrain feature.

Perhaps the most significant variations in the wind happen on the smallest levels—from second to second and foot to foot. The variation in wind flow on small scales is known as turbulence. An example of turbulence is seen when cream is poured into coffee and the cream swirls, curls and twists. Smoke rising from a fire streams upward and breaks into whirls and eddies. The exact patterns of these swirls and eddies is difficult to describe, much less predict. Eddies and swirls distinguish turbulent from non-turbulent flow.

Eddies occur when a stream stalls against its boundaries or against another stream. The stream breaks into pieces that roll over on themselves. At the boundary, the flow of the stream has zero velocity, which is why particles of dust can ride on the blade of a fan without being blown off, and why you cannot blow fine pieces of dust from the surface of a table, only the large pieces that stick up into the fast-moving breeze. At increasing distances from the boundary, the flow moves at higher velocities; and the differences in the rate of flow causes the stream to trip over itself, to curl around on itself just as a wave curls rushing up to the beach (Figure 2-5).

The variability caused by turbulence means the wind speed can change significantly in a short period of time or distance. Gusting wind can go from 10 to 30 mph in less than one second, or blow at 30 mph, 70 ft off the ground and only 15 mph at 60 ft.

Wind speed is influenced by terrain, barriers and even the roughness of the surface the wind flows over. Each of these factors, in specific situations, determines the direction and strength of the wind. Barriers or obstructions such as buildings, trees, or hills affect the flow of the wind. In Figure 2-6, it is clear that the wind flow is disturbed over a much greater area than the simple height of the obstruction.

Simply flowing over the surface of the earth affects the speed and turbulence of the wind. Generally, the wind speed tends to increase with height—the futher away the flow is from the surface. This effect is known as the *vertical wind gradient* (Figure 2-7).

Changes in the type of surface the wind flows over affects the speed of the wind, particularly at different heights. The wind moving from a smooth surface, like a body of water, to a rougher surface, like a field of high grass, exhibits behavior similar to wind passing over an obstruction. The height at which the vertical wind gradient is affected is known as the *transition height* (Figure 2-8).

The area below the transition height represents disturbed flow which can be faster, or slower and more turbulent than the normal vertical wind gradient. Though not as severe as the effects of an obstruction like a building or tree, the transition height can significantly affect the operation of a wind machine. The effects of changes in surface roughness on the transition height, and therefore the wind speed can be estimated from the chart in Figure 2-9.

Unlike obstructions and changes in the surface roughness, certain land forms accelerate the flow of the wind. As the diagram in Figure 2-10 shows, there can be a significant increase in wind speed over a ridge or hill. In this case, the crest of the ridge may have wind that is twice as strong as nearby wind. The diagram in Figure 2-11 shows the cross-sectional shapes of several ridges, and ranks them by the amount of acceleration they produce. The triangular-shaped ridge causes the greatest acceleration; the rounded ridge is a close second.

Certain slopes, primarily the few hundred yards nearest to the summit (this portion of the ridge has the

THE WIND AS A SOURCE OF ENERGY 11

Fig. 2-5. *Starry Night* by Vincent Van Gogh.

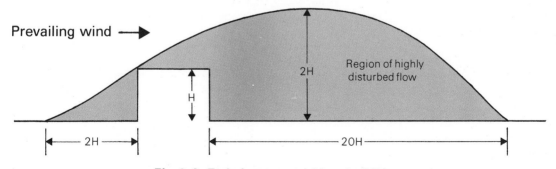

Fig. 2-6. Turbulence caused by a building.

12 YOUR WIND DRIVEN GENERATOR

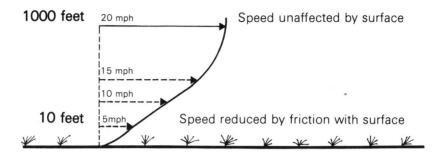

Fig. 2-7. Vertical wind gradient.

greatest influence in the wind profile immediately above the summit), increase the wind more effectively than others (Table 2-1).

Another important way to define wind is by it's energy content (Figure 2-12). When a hurricane or tornado lifts the roof off a house, it is clear that there is a lot of energy potential in the wind, but just how much energy is there in a normal breeze? Any object in motion, like the wind, contains an amount of energy based on its motion and weight. This is kinetic energy. From physics we know that kinetic energy is equal to half the mass or weight of the object times the square of its speed (V^2). The mass of wind is determined by its density (weight per unit of volume), its area, and its speed. From these factors, it can be proven that the power in the wind is a function of the cube of the velocity: V^2 times $V = V^3$.

This "cube" relationship means that small changes in wind speed result in large changes in energy content, and thus on the power delivered by a wind machine. This is perhaps the single most important concept when considering the use of wind energy. For a given amount of wind intercepted by a wind machine (deter-

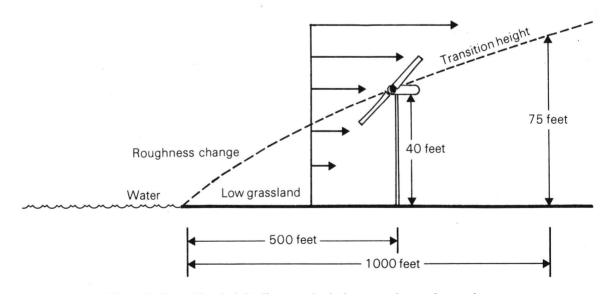

Fig. 2-8. Transition height diagram depicting one change in roughness.

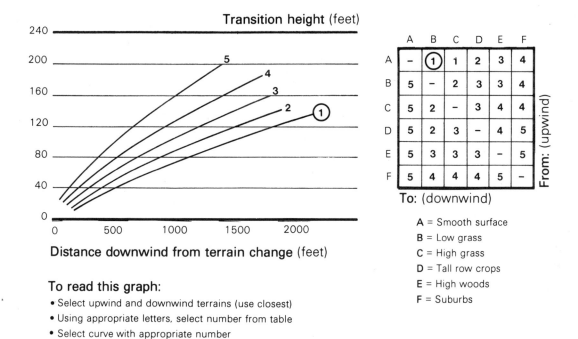

Fig. 2-9. Transition height due to change in roughness.

mined by the size of the rotor and therefore the area of wind intercepted) if the wind speed doubles the machine will produce eight times as much power:

$$2V^3 = 2^3V = 8V$$

The significance of this relationship can be seen in Table 2-2. An increase in wind speed of only two miles per hour, from 10 to 12 mph, represents a 78% increase in power.

Even with this tremendous increase in the energy content of the wind available with increasing wind speeds, it remains very dilute. The average power densities available in the United States range from 100 to 500 watts per square meter (w/m²) of intercepted area (Figure 2-13).

Unlike conventional fuels which are not being replaced as they are being consumed, wind energy is inexhaustible. The winds are a product of the sun's even heating of the earth's surface. The movement of huge

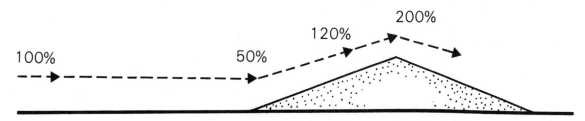

Fig. 2-10. Percent variation in wind speed over idealized ridge.

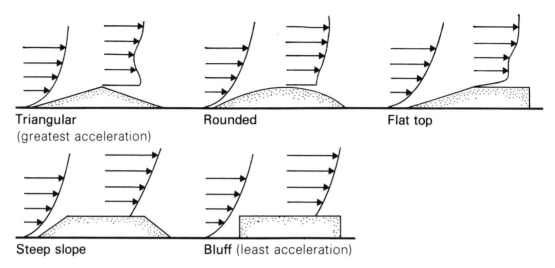

Fig. 2-11. Ranking of ridge shapes by acceleration.

air masses is caused by the greater heating of the earth's surface near the equator than the cool poles. Uneven heat distribution causes pressure differentials that result in the generation of winds. Cold polar air sinks and moves towards the equator while hot tropical air rises and moves in the upper atmosphere towards the poles. The atmosphere is attempting to obtain equilibrium of heat distribution around the earth. Global air circulation is affected by the rotation of the earth—wind circulates in a counterclockwise direction in the Northern Hemisphere and clockwise in the Southern Hemisphere. The seasonal changes caused by the rotation of the earth's axis also affects the strength and direction of the winds. The winds will continue to distribute heat around the earth for as long as the sun shines (Figure 2-14).

The environmental impact of wind energy is minimal. Because wind machines do not burn fuel, there are no exhaust fumes or air emissions. This is important in a country whose skies are clouded with smog and for an individual who wants to have his own power source in the back yard.

Each wind machine requires a certain amount of land, and some experts believe this may limit the ultimate contribution wind energy will make to national energy supplies. Land area is a major concern for wind farms—hundreds of machines spread out over hundreds of acres. Many of the machines erected in the

Table 2-1. Site Suitability Based on Slope of a Ridge.

WIND MACHINE SITE SUITABILITY	SLOPE OF THE HILL NEAR THE SUMMIT	
	PERCENT GRADE*	SLOPE ANGLE
ideal	29	16°
very good	17	10°
good	10	6°
fair	5	3°
avoid	less than 5	less than 3°
	greater than 50	greater than 27°

* Percent grade as used above is the number of feet of rise per 100 ft horizontal distance

Energy in the Wind

The energy in the wind comes from kinetic energy—the product of the mass or weight of the wind times its velocity squared (velocity²). The mass of the wind is determined by its volume. In the case of a horizontal axis machine whose swept area is a circle, this is the swept area times the wind speed:

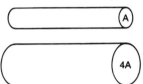

Mass × Velocity² = Energy

Once determined, the mass of this cylinder is then multiplied by the square of the speed at which it is moving. The result is that power in the wind is a function of the cube of the velocity (wind speed³) and the square of the radius of the wind machine (blade length²).

The effect of doubling the wind speed on the energy available can be portrayed graphically:

Doubling the rotor diameter has the following effect:

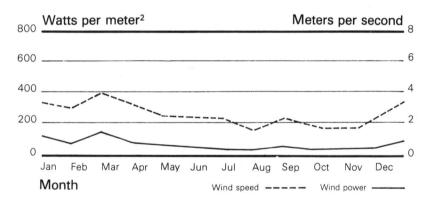

Fig. 2-12. Energy content in the wind.

Table 2-2. Percentage Change in Available Power with Changes in Wind Speed

SPEED (MPH)	PERCENT CHANGE IN POWER
5	−88
6	−78
7	−66
8	−41
9	−27
10	0
11	+33
12	+78
13	+120
14	+174
15	+238

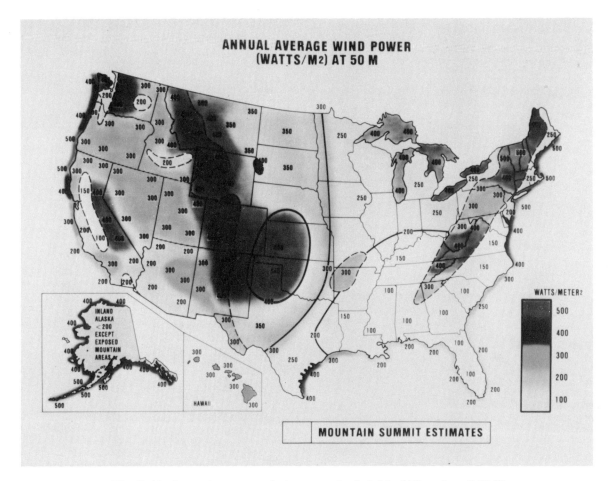

Fig. 2-13. Annual average wind power at a height of 50 meters (162 ft).

future are likely to be small, not requiring much land. Even on wind farms, multiple use is possible. Cattle can graze safely at the base of the wind machines.

Large wind machines, with rotors several hundred feet in diameter, have caused television interference up to a distance of one mile. Small units usually do not cause TV interference because the area swept by the blades is so small, and frequently their blades are nonmetallic. The proliferation of cable television, remote siting of large machines, and the use of nonmetal blades will tend to mitigate the problem.

Some wind machines are noisy. Neighbors may notice the whirring of the blades, or the creaking of a gear box running without oil. Again, the history of the automobile is a good model for the evolution of wind technology. When cars first traveled the roads of America, they terrified animals and people alike with gunshot backfires and the rumble of poorly muffled motors. But as the technology became more sophisticated, the public accepted the car as a part of modern life.

Finally, local zoning boards may contend that a wind machine will disrupt the beauty of the community. This is a difficult issue because what is objectionable to one person may be beautiful to another. Technology is not inherently ugly, as the recent trend of "high-tech" interior design and architecture proves. A carefully

THE WIND AS A SOURCE OF ENERGY 17

The direction and intensity of world winds are affected by the earth's rotation, the proportion of land mass to sea mass, and mountain ranges. The development of ocean trade routes depended on the reliability of *trade winds*, caused by the spreading out of sinking air over the horse latitudes in both southerly and northerly directions. In the temperate zones, warm tropical air mixes with polar air producing violent and varied weather, a climate favorable to agriculture.

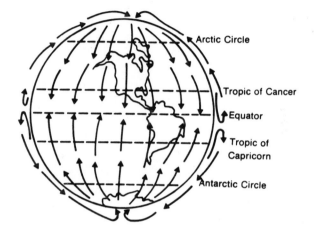

Directions of winds if the earth did not rotate. Cold polar air moves toward the equator as warm tropical air rises and moves toward the poles.

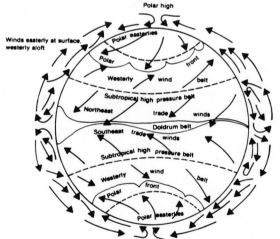

As the air over the equatorial region rises to the top of the troposphere (as high as 60,000 feet), the surface equatorial region is characterized by hot, sultry air with little wind activity except those locally generated. The name of this region, *the doldrums*, has become synonymous with the listlessnesss and general lack of activity.

Fig. 2-14. Global wind patterns.

designed machine will stand in harmony with its environment, and represent a sense of values that is likely to be appreciated by the community (Figure 2-15).

The environmental concerns—land use, television interference, noise, and aesthetics—appear now to be the only potential environmental problems of wind machines. They stand in contrast to the effects of conventional fuels: smog, radioactive waste, acid rain, and melting polar icecaps.

Fig. 2-15. Modern wind machines.

3
IS WIND ENERGY PRACTICAL?

Determining whether wind energy is practical for a specific application involves answering three more specific questions: how much energy is needed, how much wind is available, and can a wind machine convert the available wind into the energy needed?

History proves that it is possible to tap the energy in the wind, and recent analyses suggest that the potential is quite large. But can the wind deliver the quantity and quality of energy to meet the specific needs of an individual?

HOW MUCH ENERGY IS REQUIRED?

The practicality of wind energy is determined by economics. Machines are available from a few hundred watts to thousands of kilowatts in electrical capacity, and it is likely that there is a machine that will meet all your energy needs. But because conservation is the most immediate and economical way to reduce energy bills, because the utility will not pay as much for the energy you sell them as they charge for the energy they sell you, and because larger wind machines cost more, an optimum wind system will produce only as much energy as you need, when you need it.

If you are considering wind energy for an application where electricity is already used, previous utility bills will determine how much energy has been used in the past. Utility bills show kilowatt hours consumed for the billing period, usually one month. The utility frequently keeps extensive records on consumption patterns in their service area, by class of customer—residential, farm, or commercial.

Single family residences can be divided into three groups, based on their requirements for electricity:

- lighting and appliances only;
- hot water, lighting and appliances;
- space heating, hot water, lighting and appliances.

The total energy required and the amount of energy needed for lights, appliances, hot water or space heating varies by house and by region. Space heating requirements depend on the size, construction, and insulation of the house; the living patterns of the residents; and the weather in the region. The water heating requirements are determined primarily by the number of occupants and how the appliances that use hot water, like

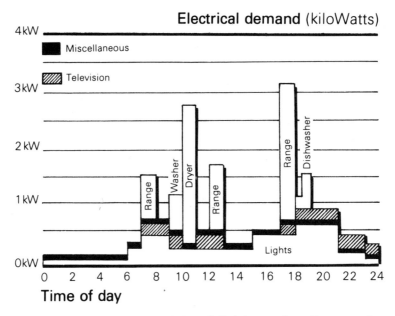

Fig. 3-1. Typical residential load: lighting and appliances only.

the dishwasher and clotheswasher, are used. The electricity required for lights and appliances is determined by the size of the house, the number and habits of the residents, and the range of appliances. For a house that uses a range, refrigerator, dryer, and dishwasher, the daily consumption pattern can be represented by the diagram in Figure 3-1.

This load pattern is not particularly favorable for wind machines because it is uneven with small constant demand over 24 hours. The peak demand for this load is 3.2 kW and the load factor, (daily consumption ÷ peak demand) × 24 hours is 0.26. These appliances have little flexibility to accomodate windy or calm periods.

The addition of a hot water load to that of lights and appliances improves the wind machine output load match (Figure 3-2). The hot water tank provides 5-10 kWh equivalent of energy storage and provides the user with opportunities to control the electrical demand. The load pattern for hot water, lights, and appliances has a more steady character, a significant constant demand over 24 hours, and a higher load factor (0.46) than the lights and appliances only pattern. Between 0.75 and 1.5 kW of demand exist for more than 20 hours per day, allowing nighttime wind machine output to be consumed directly on the site, rather than being fed back to the utility.

The all-electric residence has the largest typical residential electrical load. The annual energy consumption for heating alone ranges from 19,000 to 56,000 kWh depending on the climate and other characteristics. The seasonal load profile of a typical, well-insulated, 1600 ft^2, two-story house is shown in the series of diagrams in Figures 3-3 to 3-7.

Understanding and assessing your specific electrical energy needs requires not only knowing how much electricity you use over the year, but the monthly and even hourly variations as well. These variations can have a substantial impact on the characteristics and economics of the optimum machine for your application.

IS WIND ENERGY PRACTICAL? 21

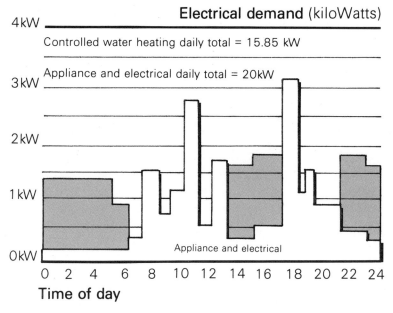

Fig. 3-2. Residential load: hot water, lighting, and appliances.

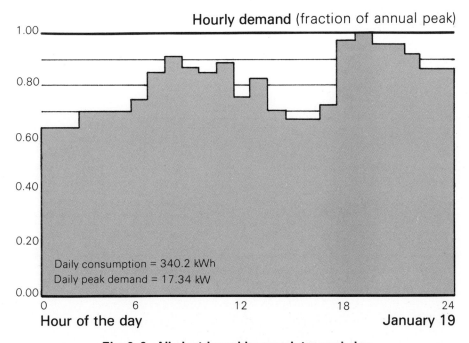

Fig. 3-3. All electric residence: winter peak day.

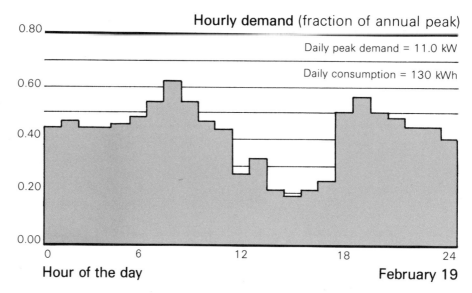

Fig. 3-4. All electric residence: typical winter day.

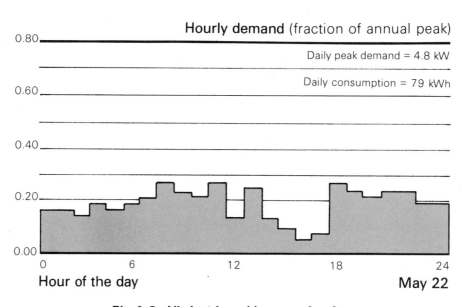

Fig. 3-5. All electric residence: spring day.

IS WIND ENERGY PRACTICAL? 23

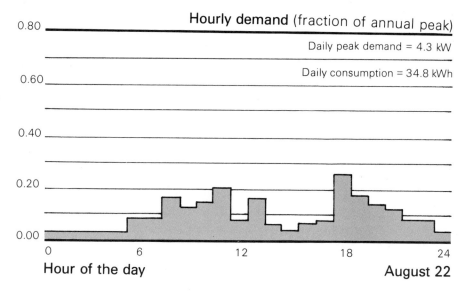

Fig. 3-6. All electric residence: summer day.

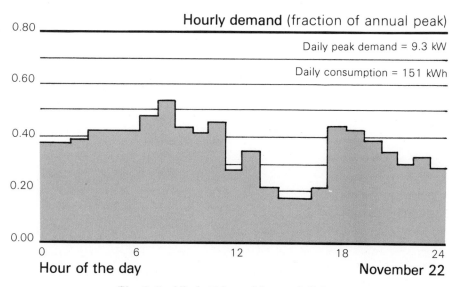

Fig. 3-7. All electric residence: fall day.

HOW MUCH WIND IS AVAILABLE?

The wind can be used successfully and economically in many parts of the country. Once an estimate of the energy requirements in kilowatt-hours per month is made and the consumption patterns charted, the focus shifts to the amount of wind that is available.

The broadest description of the amount of wind available is annual average wind speed. But, like annual consumption, it only allow you to guess whether there is enough wind to make wind energy practical. There is a wide variability of wind speeds from month to month. Consider the data for Cheyenne, Wyoming in Table 3-1.

Because of the cube relationship between wind speed and available energy, there is 250% more power available in December than August. This chart demonstrates how crude a figure like annual average wind speed really is. However, if this data represented a site being considered, wind energy could be employed economically, even though there may be little power produced in July, August, or September.

The best single source for more detailed wind data, including monthly and hourly variations, is the Wind Resource Atlas prepared by Battelle Pacific Northwest Laboratories for the Department of Energy. There are twelve volumes for various regions in the United States, Hawaii, Alaska, and Puerto Rico (Figure 3-8 and Table 3-2).

Each atlas includes introductory and background information, a regional summary of the wind resource, and assessments of each state in the region. The introductory material describes how the assessment was developed and how it should be interpreted. The wind resource is discussed on a regional scale, and the results for each state in the region are presented in separate chapters. Each state wind resource is described in much greater detail than the regional resource, and the features of the data collection stations used in the assessment are discussed (Figure 3-9).

The first figure in each state chapter shows the major geographical (mountains, rivers) and man-made (cities, towns) features in each state. The second figure portrays the topography of the state in shaded relief. This map allows the reader to visualize the terrain surrounding a potential site. Superimposed on the state maps is a grid of dashed lines, one-third degree in longitude (approximately 20 miles) by one-quarter degree

Table 3-1. Monthly Average Wind Speeds—Cheyenne, Wyoming.

MONTH	WIND SPEED (MPH AVERAGE)
January	15.9
February	16.3
March	15.3
April	14.2
May	12.6
June	11.6
July	10.0
August	9.9
September	10.9
October	11.2
November	13.8
December	15.4
annual average:	13.1

IS WIND ENERGY PRACTICAL? 25

Fig. 3-8. Regions of the *Wind Resource Atlas.*

in latitude (approximately 15 miles). This grid is repeated in all subsequent maps to provide a frame of reference for locating the same place on different maps.

A land surface form map is also provided for each state (Fig. 3-10). How the land surface form identifies terrain features considered to have good exposure to the wind is discussed. To interpret wind power maps, it is crucial to understand what constitutes well-exposed terrain.

Later figures identify and locate wind data sites used in preparing the atlas. All locations in each state where the National Climatic Center (NCC) has wind data are shown. However, these atlases are based on data from other sources—power plants, fire watch towers, military bases, as well as NCC data. The integration of all these data sources is, in part, what makes the atlases the best current source of wind data.

The wind energy resource in each state is illustrated using a series of maps. The wind power maps represent a careful synthesis of the available wind data, guided by what is known about the flow of air near the earth's surface. The wind data is transformed into the map of average annual wind power densities at typically exposed sites. Wind power density was chosen for describing the wind power, rather than mean wind speed, because the combined effects of the distribution of wind speeds, the dependence of power density on air density, and the cube relationship between power and wind speed can all be reflected in wind power density (w/m²) but not mean wind speed (Table 3-3).

A certainty rating provides a measure of the ability to objectively evaluate the wind resource in each cell of

Table 3-2. Volumes of the *Wind Resource Atlas*.

PUBLICATION #		VOLUME, AREAS COVERED	PRICE
PNL-3195	WERA-1	Vol. 1—The Northwest Region (Idaho, Montana, Oregon, Washington, Wyoming)	$6.50
PNL-3195	WERA-2	Vol. 2—The North Central Region (Iowa, Minnesota, Nebraska, North Dakota and South Dakota)	$6.00
PNL-3195	WERA-3	Vol. 3—The Great Lakes Region (Illinois, Indiana, Michigan, Ohio, and Wisconsin)	$6.00
PNL-3195	WERA-4	Vol. 4—The Northeast Region (Connecticut, Maine, Massachusetts, New Hampshire, New Jersey, New York, Pennsylvania, Rhode Island, Vermont)	$6.50
PNL-3195	WERA-5	Vol. 5—The East Central Region (Delaware, Kentucky, Maryland, North Carolina, Tennessee, Virginia, West Virginia)	$6.50
PNL-3195	WERA-6	Vol. 6—The Southeast Region (Alabama, Florida, Georgia, Mississippi, South Carolina)	$6.00
PNL-3195	WERA-7	Vol. 7—The South Central Region (Arkansas, Kansas, Louisiana, Missouri, Oklahoma, Texas)	$6.50
PNL-3195	WERA-8	Vol. 8—The Southern Rocky Mountain Region (Arizona, Colorado, New Mexico, Utah)	$6.00
PNL-3195	WERA-9	Vol. 9—The Southwest Region (California, Nevada)	$5.00
PNL-3195	WERA-10	Vol. 10—Alaska	$6.00
PNL-3195	WERA-11	Vol. 11—Hawaii and Pacific Trust Territories	$4.75
PNL-3195	WERA-12	Vol. 12—Puerto Rico and U.S. Virgin Islands	$4.50

These volumes are available from the U.S. Government Printing Office:

U.S. Government Printing Office
Superintendent of Documents
Washington, DC 20402

the grid imposed on the state maps. The degree of certainty or confidence in the atlas's description of the wind power depends on the quality and quantity of data, the complexity of terrain, and variations in wind over short distances (Figure 3-11).

The annual average wind power density classes compress into a single number, time varying trends on several scales—annual, seasonal, monthly, and daily. In the atlas, graphs are provided that show how wind speeds and wind power varies from year to year, month to month, hour to hour (Figures 3-12 to 3-15). These graphs are based on existing NCC weather stations where data is available. Considering information at one of these sites as representative of some other location must be tempered with the realization that wind characteristics are extremely site dependent. The degree of correlation from site to site depends on the

Table 3-3. Classes of Wind Power Density at 10 meters and 50 meters.

CLASS	WIND POWER (W/M^2) @ 10M (33 FT.)	WIND SPEED M/S (MPH)	WIND POWER (W/M^2) @ 50M (164 FT.)	WIND SPEED M/S (MPH)
1	100	4.4(9.8)	200	5.6(12.5)
2	150	5.1(11.5)	300	6.4(14.3)
3	200	5.6(12.5)	400	7.0(15.7)
4	250	6.0(13.4)	500	7.5(15.7)
5	300	6.4(14.3)	600	8.0(17.9)
6	400	7.0(15.7)	800	8.8(19.7)
7	1000	9.4(21.1)	2000	11.9(26.6)

IS WIND ENERGY PRACTICAL? 27

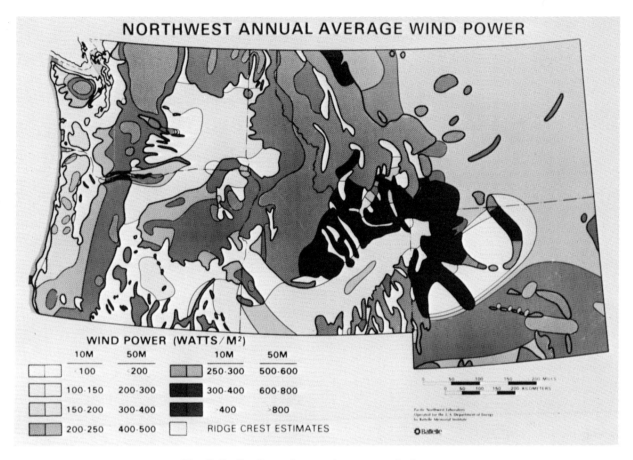

Fig. 3-9. Northwest annual average wind power.

similarity of the topography, the weather patterns that affect the sites, and the obstruction to the wind in the vicinity of the site.

Although the Wind Resource Atlas is the most complete compilation of wind energy data available, it is a crude tool for estimating the wind power potential of a particular site. After reviewing the Atlas, a number of problems may remain. The 'cell' may have a very low certainty rating. The terrain may be complex, and increases in the wind speed may be available because of terrain features. The site may appear marginal based on the data in the atlas, but circumstances may lead you to believe wind energy could be employed economically.

One method of estimating wind power potential at a site that is gaining popularity is to inspect local vegetation, particularly coniferous trees, for deformation due to the wind (Figure 3-16). Trees were first used as indicators of wind power potential by Palmer Putnam on a large wind generator project in Vermont in the 1940s. He described an index that relates tree deformation due to wind speed. Recent research shows that this can be a reliable indicator of wind power potential. There are a number of practical limitations to the use of trees to determine wind speeds. While flagged trees may indicate high wind speeds, trees that are not flagged do not necessarily mean that the winds are light. There may be locations where the wind comes

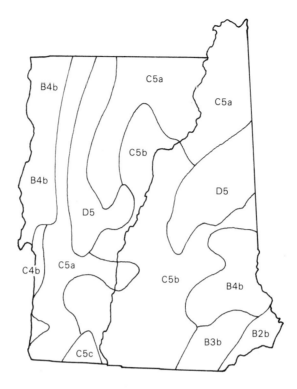

Fig. 3-10. Land surface form legend.

PLAINS

A1 flat plains
A2 smooth plains
B1 irregular plains, slight relief
B2 irregular plains

PLAINS WITH HILLS OR MOUNTAINS

A,B3a,b   plains with hills
B4,a,b   plains with high hills
B5,a,b   plains with low mountains
B6,a,b   plains with high mountains

OPEN HILLS AND MOUNTAINS

C2 open low hills
C3 open hills
C4 open high hills
C5 open low mountains
C6 open high mountains

HILLS AND MOUNTAINS

D3 hills
D4 high hills
D5 low mountains
D6 high mountains

TABLELANDS

B3,c,d   tablelands, moderate relief
B4,c,d   tablelands, considerable relief
B5,c,d   tablelands, high relief
B6,c,d   tablelands, very high relief

SCHEME OF CLASSIFICATION

Slope (1st letter)

A more than 80% gently sloping
B 50-80% gently sloping
C 20-50% gently sloping
D less than 20% gently sloping

Local Relief (2nd letter)

1   0-30 m (1-100 ft.)
2   30-90 m (100-300 ft.)
3   90-150 m (300-500 ft.)
4   150-300 m (500-1000 ft.)
5   300-900 m (1000-3000 ft.)
6   900-1500 m (3000-5000 ft.)

Profile Type (3rd letter)

a   more than 75% of slope is in lowland
b   50-75% of slope is in lowland
c   50-75% of slope is on upland
d   more than 75% of slope is on upland

IS WIND ENERGY PRACTICAL? 29

Fig. 3-11. Certainty rating map.

Rating   *Definition*

1   Lowest degree of certainty. A combination of the following conditions exist:
       1) no data exists in the vicinity of the cell
       2) the terrain is highly complex
       3) various meteorological and topographical indicators suggest a high level of variability of resource
2   A low-intermediate degree of certainty.
       1) little or no data exist, but the small variability of resource and low complexity of terrain suggest that the wind resource will not differ substantially for the resource in nearby areas.
       2) Limited data exists, but the terrain is highly complex
3   A high-intermediate degree of certainty.
       1) there are limited wind data, but the terrain is not complex.
       2) considerable wind data exist but in moderately complex terrain, or where a moderate variability of the resource is likely to occur.
4   The highest degree of certainty. Quantitative data exist at exposed sites in the vicinity of the cell and can be confidently applied to exposed areas because of the low complexity of the terrain.

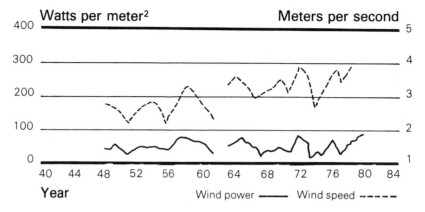

Fig. 3-12. Interannual wind power.

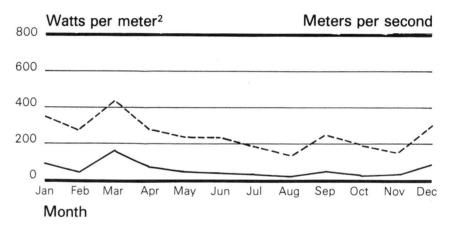

Fig. 3-13. Monthly average wind power.

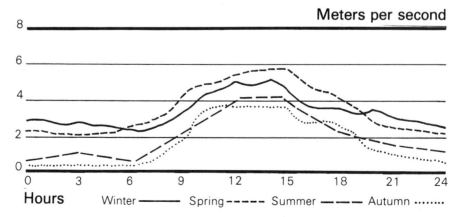

Fig. 3-14. Diurnal wind speed by season.

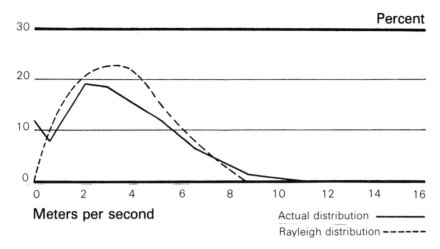

Fig. 3-15. Annual average wind speed frequency.

from several directions, and the persistence of wind from any one is not enough to cause flagging. Also, particularly at lower wind speeds, the flagging may be slight, and hard to detect.

In spite of the possible errors that are inherent in using trees to estimate wind power potential, they can provide a simple, inexpensive guide for rating an area's wind power potential, indicating patterns of wind flow over complex terrain, and verifying data from other sources.

Using the Wind Resource Atlas, it will be possible to make a basic determination of the wind energy available at a particular site. There are many factors other than wind speed that affect the practicality of a wind machine installation. A rule of thumb is that a particular site should have the potential of a 12 mph annual average wind speed to be considered economically viable where conventional utility power is available.

WHAT SIZE MACHINE?

Once the energy demands have been determined and the amount of wind energy available has been estimated, the next step in deciding whether wind energy is practical is to consider the wind equipment itself—the question of selecting the right size for a particular wind regime. Although it is common practice to describe a wind machine in terms of the rated electrical output at a specific wind speed, 5 kilowatts at 25 miles per hour for example, it is more accurate to place primary emphasis on the diameter of the rotor. This is more critical than the size of the generator because it defines how much power is available to be converted into electricity.

The significance can be seen by comparing two wind machines. The first is rated at 6 kilowatts in a 27 mph wind and has a rotor diameter of 5 meters (16.4 ft.). The second machine is rated at 4 kilowatts in a 16 mph wind and has rotor diameter of 10 meters (32.7 ft.). Assuming a 12 mph average monthly wind speed to calculate the output of the two machines, it can be seen that the first machine will produce approximately 400 kilowatt-hours per month, which the second machine will produce over 1,000 kilowatt-hours per month. In spite of the fact that the first has an electrical rating 50 percent greater than the second, the significantly larger rotor diameter produces the rated output of the generator at a much lower, and therefore

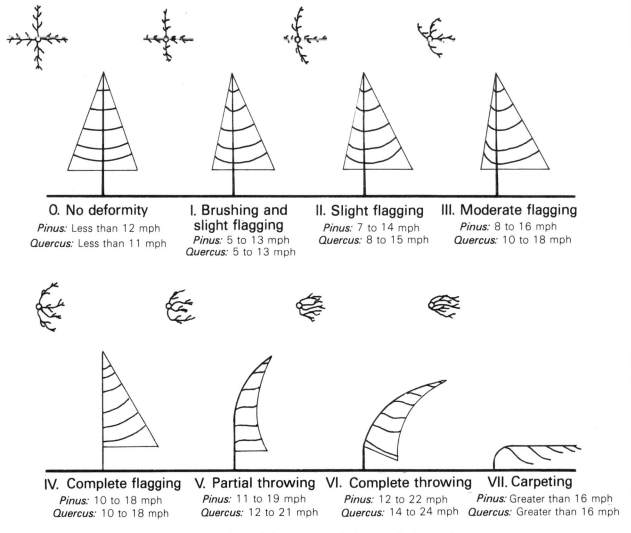

Fig. 3-16. Tree deformation relating to wind speed.

more frequent wind speed, and the result is over twice as much energy. The simple graph in Figure 3-17 shows typical annual energy outputs for wind machines of various diameters at a range of annual average wind speeds.

The energy actually produced at a specific site can vary by as much as 20 percent from the output represented by this graph. This is due to widely varying monthly wind speeds and the operating characteristics of different machines. Beyond a certain average wind speed, the energy production does not increase. This is because with very high average wind speeds, the machine will probably be operating at its peak capacity much of the time. The annual energy output curves tend to flatten out earlier for smaller machines.

The second important characteristic, after rotor diameter, in selecting an approximate size machine is also related to energy output. Although it is common to think of the electrical demand in terms of power,

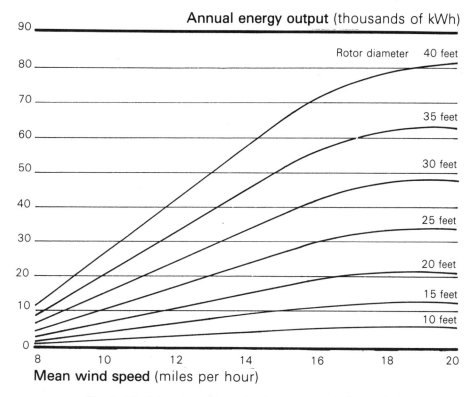

Fig. 3-17. Annual energy output vs. mean wind speed.

"100 amp service" for example, wind machines should be considered in terms of the energy they produce rather than the electrical capacity. Think about the kilowatt-hours a machine will deliver, rather than its kilowatt capacity rating. The manufacturer or distributor should provide a graph, similar to the one in Figure 3-18, showing the kilowatt-hours produced over time at various average wind speeds.

For a specific installation, it may be possible to determine an optimum machine capacity based on minimizing the total cost of electrical energy over the operating life of the machine. Total cost is the cost of electricity from the wind machine, plus the cost of backup power purchased from a utility. The most important influences on the selection of size are:

- Available wind resource
- Power requirements of user
- The cost of electricity from conventional sources
- The rate the utility will pay for excess electricity returned to the utility network

In many cases no optimum size can be determined. However when the size is based on minimizing the total electrical costs, the sizes can vary considerably as a function of the factors listed above. The selection of a specific size machine may be based on factors other than minimizing total cost of electrical energy. For example, there may simply be an upper limit of what you are willing to pay for a wind system, or the utility

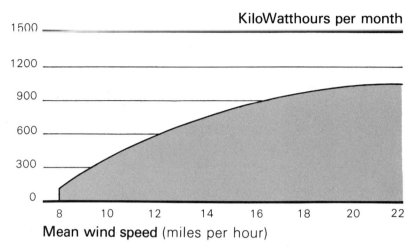

Fig. 3-18. Typical wind machine energy production.

'buy-back' rate is so low, the optimum size will be determined by setting an upper limit on the amount of electricity sold to the utility.

Determining your energy requirements, assessing the available wind resource, and selecting the 'optimum' size machine are the basic tools for determining whether wind energy is practical.

4
CHOOSING THE RIGHT WIND MACHINE

Interpreting specification sheets; test results; predicting energy production with the Rayleigh distribution and the power curve; available standards; talking with other users, and the use of consultants.

Choosing the right wind machine can only come after you collect accurate information on how much and when you use electricity and have determined how much wind is available at the site and the approximate size wind machine you need. The decision is complicated, involving a broad range of issues, and there is no single correct answer. However, there are a number of steps the potential wind machine owner can take to ensure that the machine selected will be appropriate for the application and represent a good investment.

INTERPRETING SPECIFICATION SHEETS

The first information a potential wind machine owner is likely to obtain is a brochure from the manufacturer or distributor extolling the virtues of their machine. The quality of these brochures range from extremely detailed and useful booklets to vague and even misleading one-page flyers. Be cautious if the most basic parameters of the machine are not described. Typically, specification sheets include two basic types of information: design specifications and performance statistics (Figure 4-1). The design specifications include the physical characteristics of the specific machine as well as major operating characteristics. This includes:

- Design output: kilowatts at a specific wind speed
- Rotor speed control: description of overspeed controls
- Operating wind speeds
  cut-in wind speed: the wind speed when the machine begins producing power
  cut-out speed: the wind speed above which no power production is expected
- Rotor configuration
  diameter: describes the area wind intercepted
  rotor type: configuration; horizontal axis or vertical axis, upwind or downwind of tower
  number of blades
  blade material: wood, steel, plastic, etc.
- Machine description: description of the machine's operation and features.

SPECS · HR2

General

Rotor Configuration:	Horizontal axis, upwind, 3-bladed
Power Output (rated):	2200 watts at 9 m/s (20 mph)
Mean Power Output (MPO)*	500 w @ $\overline{V}$ = 10mph(4.5 m/s)
	760 w @ $\overline{V}$ = 12mph(5.4 m/s)
	1150w @ $\overline{V}$ = 16mph(7.2 m/s)
Voltage (nominal):	24, 32, 48, 110 vdc
Interface Requirement:	Battery storage
Transmission:	None required (direct drive)
Yaw Control:	None required (free yawing)
System Weight (excluding tower):	356 kg (785 lb)
Tower Height (minimum):	12 m (40 ft)
Tower Weight (12 m):	545 kg (1200 lb)

Wind Turbine

Rotor Diameter:	5 m (16.4 ft)
Blade Material:	Wood composite
Cut-in Wind Speed:	3.6 m/s (8 mph)
Rated Wind Speed:	9 m/s (20 mph)
Speed Control Initiation:	9.3 m/s (21 mph)
System Shutdown:	47 m/s (105 mph)
Axial Thrust (maximum):	2800 N (630 lb)
Overspeed Control:	Variable Axis Rotor Control System (VARCS)

Electrical System

Generator Type:	3 phase, synchronous alternator with wound stator
Field Configuration:	Lundel type, shunt-connected
Rated Output:	2200 watts at 250 rpm
Maximum Output:	3000 watts
Rectification:	Silicon diode full-wave bridge
Voltage Regulation & Battery Protection:	Solid state field control with over-voltage protection

Environmental Conditions

Temperature:	-60°C to 60°C (-70°F to 140°F)
Wind, steady:	54 m/s (120 mph)
Wind, gusting:	75 m/s (165 mph)
Rain, dust, industrial atmosphere, salt water spray	Sealed construction; weathertight fittings and connectors; corrosion resistant materials

*The MPO is a calculation of the average amount of power the SWECS will deliver to the batteries. It is based on assumed wind patterns; your actual performance may vary.

Fig. 4-1. Typical manufacturer's specification sheet.

As can be seen from the example on the following page, some manufacturer's specifications go well beyond this basic list. In the case of some machines, the National Small Wind Systems Test Center at Rocky Flats, Colorado publishes "Performance Summary Sheets" of machines they have tested. An example is shown in Figure 4-2.

Some of the information typically provided in manufacturer's specification is more useful than others. We can look at each element and consider its impact on selecting the 'right' machine.

Design Output

The design output is the power output at which the machine is rated, and the lowest wind speed at which this output is predicted. It is also known as *rated output*. Though widely used, this method of describing a wind machine is seriously flawed. The most significant problems are:

1) Rated output tells the potential purchaser nothing about the energy (kilowatt hours) production of a machine. For example, a machine rated at a lower capacity that delivers the capacity at a lower wind speed is likely to produce much more power than a machine with a higher rated capacity and a smaller rotor diameter.
2) The rated output can be different than the maximum capacity of the machine. That is, a machine could be rated at 2 kW, but capable of producing 3 kW in a stronger wind.
3) The rated output can mislead. A machine rated at 10 kW in a 30 mph wind may not be larger than a machine rated at 5 kW in a 20 mph wind, although at first glance, it may appear that the 10 kW machine is larger, or costs less on a dollars per kilowatt capacity basis.

Although this method is inadequate for assessing a machine, or comparing two machines, its wide use suggests that it will continue to be employed.

Rotor speed control

The rotor speed control is the method used to prevent excessive rotational speed. This protects the wind machine from structural damage that may result from high winds. This is an extremely critical feature and is frequently the "core" of a wind machine's design. Its success or failure determines the success or failure of the whole system. For the potential wind machine owner, assessing this critical system is not an easy task. It cannot be reduced to a single number, and is frequently provided only as a narrative description. There are many approaches to rotor overspeed controls. Most manufacturers have different design philosophies, and each feels that their approach is best. Until there is a great deal more operating experience, there is no clear-cut best type of rotor speed control.

Basic machine design principles can be used to assess this critical system and the different approaches that are taken. Three design principles are *simplicity, redundancy,* and *precision.* Simplicity means minimizing the number of parts which generally increases system reliability. Redundancy means having backup systems available if the primary system fails, which also tends to increase reliable operation. Reliability is a key element in a system where a failure is likely to be catastrophic. Finally, precision in the design should indicate that the forces which initiate control action are a direct result of the destructive forces on the machine, and that the control response directly alleviates these destructive forces. Violating design prin-

AUGUST 1980

Rocky Flats Performance Summary

SENCENBAUGH Model 1000-14

MANUFACTURER'S SPECIFICATIONS
(See U.S. CONTACT for available options)

DESIGN OUTPUT:
    1.0 kW @ 10.3 m/s (23 mph)

ROTOR SPEED CONTROL:
    Mechanical, rotor turns edgewise to wind, activated by excessive thrust load.

OPERATING WIND SPEEDS:
    CUT-IN: 2.7 m/s (6 mph)
    CUT-OUT: 27 m/s (60 mph)

ROTOR CONFIGURATION:
    ROTOR DIAMETER: 3.65 m (12 ft)
    ROTOR TYPE: Horizontal axis, upwind, fixed-pitch.
    NUMBER OF BLADES: 3
    MATERIAL: Wood (Sitka spruce), bonded copper leading edge, epoxy finish.

GENERATOR/TRANSMISSION:
    OUTPUT: 3-∅, 6-pole alternator, rect. to dc.
    GEARBOX: Helical, 3:1 ratio

MACHINE DESCRIPTION:

The Sencenbaugh is a lightweight machine manufactured by Sencenbaugh Wind Electric, Palo Alto, California. The blades are manufactured from Sitka spruce, with a bonded copper leading edge and a polyurethane finish. The main generator casting is 356T6 aluminum alloy. The alternator is coupled through a 3:1 helical gearbox, and is driven by a 3-bladed propeller 3.65 m in diameter. Overspeed control is provided by using the increasing wind pressure of the propeller (and the resultant propeller thrust) to swing the alternator assembly (which is offset from the bearing support column) out of the oncoming wind. The foldable tail automatically reopens as wind speed decreases due to gravitational forces on the tail assembly. Tail offset and inclination with respect to the rotor may be varied, thus changing the cut-in speed of the machine.

U.S. CONTACT:

    SENCENBAUGH WIND ELECTRIC
    P.O. BOX 11174
    PALO ALTO, CALIFORNIA 94306

    (415) 964-1593
    JIM SENCENBAUGH

This PERFORMANCE SUMMARY was prepared and published by the Rockwell International Corporation, Energy Systems Group, Rocky Flats Plant, Wind Systems Program, P.O. Box 464, Golden, CO 80401
for the
U.S. Department of Energy, Office of Solar Power Applications
Federal Wind Energy Program
Contract DE-AC04-76DP03533

DISCLAIMER

This report was prepared as an account of work sponsored by the United States Government. Neither the United States nor the United States Department of Energy, nor any of their employees, makes any warranty, express or implied, or assumes any legal liability or responsibility for the accuracy, completeness, or usefulness of any information, apparatus, product, or process disclosed, or represents that its use would not infringe privately owned rights. Reference herein to any specific commercial product, process, or service by trade name, mark, manufacturer, or otherwise, does not necessarily constitute or imply its endorsement, recommendation, or favoring by the United States Government or any agency thereof. The views and opinions of authors expressed herein do not necessarily state or reflect those of the United States Government or any agency thereof.

08/80 dlm

Fig. 4-2. A Rocky Flats Performance summary.

CHOOSING THE RIGHT WIND MACHINE   39

ciples has made Rube Goldberg contraptions famous, but successfully employing them is much harder to characterize (Figure 4-3).

There are three basic types of rotor speed control, although there are many variations within these categories:

1) Changing the angle of attack of the blade. The angle at which the blade meets the wind can be increased or decreased, eliminating aerodynamic lift, the driving force on the rotor. Increasing the angle is known as stall and decreasing the angle is known as feather. Feathering is generally preferred to stall, because it reduces the overturning moment on the tower. The force that frequently initiates this action is the centrifugal force on the rotor that builds up as the wind speed and rotational speed increases.

2) Changes in the plane of rotation. Changing the plane of rotation can come from a folding tail that faces the blades out of the wind, or the blades themselves can be moved around a vertical or horizontal axis so that, again, they face the wind edgewise rather than "frontally".

3) Wind machine brakes can be aerodynamic (as in the case of tip flaps) or mechanical (as in the case of a disc brake). Brakes can be activated by a signal, as from an anemometer, or a force, like the centrifugal force that builds up as the speed of the blades increases.

Drawing courtesy ARCO Inc.

Fig. 4-3. Violating good design.

In comparing overspeed controls, apply the basic machine design principles of simplicity, redundancy, and precision. The best system will be one that has few moving parts, features redundancy or at least "fails safe", and deals with the destructive forces and their impact on the machine in a direct, precise manner.

Operating wind speeds

Frequently, two figures are provided under operating wind speed: the cut-in and cut-out speed. Cut-in is the wind speed at which the wind machine begins producing power. Cut-out is the speed at which the wind machine is no longer expected to produce power.

Although these figures are widely used, they are of little real value when comparing two wind machines. The cut-in wind speed has little impact on energy production both because the energy content in low winds is minimal, and because overall system efficiencies are generally poor when operating far below capacity. The cut-out wind speed also does not significantly affect energy production, because very strong winds are so rare that it is more economical to "shut down" the machine than build it strong enough to extract the extra energy. If your site has extremely high average wind (16 mph or above), you should consider a wind machine with a high cut-out wind speed that would deliver more energy. However, in machines that are not designed to "shut down" in very high winds (over 50 mph for example), the potential wind machine owner should consider whether continuous operation will result in lower overall system life (Figure 4-4).

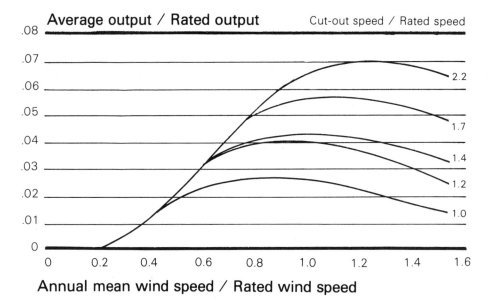

Fig. 4-4. Effects of operating wind speeds.

Rotor Configuration

Rotor configuration describes the size of the rotor, its physical configuration and the construction materials. This also includes rotor diameter, rotor type, number of blades, and blade materials.

Rotor diameter, after available wind, is the single most important factor in determining how much energy a given wind machine will produce. Because the energy available to be extracted is related to the area swept by the blades, doubling the rotor diameter makes available four times as much power. Rotor diameter is the best functional criteria for assessing the size of a wind machine rather than the more commonly used *rated output*.

Rotor type can be horizontal or vertical axis. Horizontal axis wind machines can be mounted either upwind or downwind with respect to the tower. For the potential wind machine owner, the differences between these configurations means little compared with the importance of diameter, cost, or company reputation. Likewise, the number of blades means relatively little. In the past, there has been a great deal of discussion among wind machine designers about the trade-offs involved between two or three blades, horizontal or vertical axis, and even upwind or downwind configurations. For the potential wind machine owner these questions have little bearing on whether the machine will produce enough power for the application, or whether it can reasonably be expected to last fifteen or twenty years.

The blade material is another important issue. While blade *fatigue* is probably the largest single structural issue facing the wind machine designer, and wood or plastic composites have good fatigue-resistant properties, it does not follow that only wood or plastics make good wind machine blades. Machines can be designed to use blades of aluminum, steel, or any other material; and the designer can take other steps to deal with fatigue. A rotor with fatigue-resistant blades is desirable, but other materials should not be automatically discounted.

Generator/Transmission

Generator/transmission describes the mechanism that produces the power, the type of power (alternating or direct current voltage), and the transmission, if any, used to increase the shaft rpm to meet generator requirements.

The most important aspect of the generator is the type of power that is delivered to the load. Direct current, unlike alternating current, can be stored in batteries and is best suited for remote applications. While many modern machines use alternators which produce alternating current, it is rectified (converted to direct current) internally and delivers direct current to the load. A machine that delivers direct current at the generator terminals, but is tied to the utility grid, must have a power conditioning device, usually an inverter, to be compatible with the utility grid.

The type of transmission used is of little significance to the potential wind machine owner, except that experience has shown that gears tend to perform more reliably than other types of transmission. The amount of speed increase can vary widely. Generally, the larger the diameter, the larger the speed increase required. Many designers feel that the use of a direct-drive generator or alternator which eliminates any step-up transmission, although more expensive, offers reliability advantages.

REVIEWING TEST RESULTS

After reviewing and comparing manufacturer's specification sheets, the prospective wind machine owner should obtain test data on the specific machine. One independent source of test information is the National Small Wind Energy Conversions Systems Test Center, widely known as *Rocky Flats* (Figure 4-5). The test center is operated by Rockwell International for the U.S. Department of Energy and is located in Golden, Colorado.

42 YOUR WIND DRIVEN GENERATOR

Fig. 4-5. Wind machines at Rocky Flats' test center.

The test center has approximately 30 towers for testing wind machines, and facilities for laboratory testing of equipment. The center also uses a *controlled velocity* test facility operated by the Department of Transportation in Pueblo, Colorado (Figure 4-6). In controlled velocity tests, wind machines are mounted on a moving vehicle, and propelled at various speeds to simulate the wind. This test is frequently used by manufacturers to develop *power curves* for a wind machine.

In normal testing, the wind machine operates in natural winds at the test site. The data collected at Rocky Flats is the same data a manufacturer collects during testing: power output versus wind speed; energy production over time; operating characteristics; and performance with various types of electrical loads. The data collected at Rocky Flats appears in various test reports, performance summaries, and failure reports. While much of this is very good information for the potential wind machine owner, it is important to realize that the location of the test center was selected in part because the winds are extremely strong and turbulent during some parts of the year, and that many of the machines being tested are prototypes. Based on data col-

Fig. 4-6. Controlled velocity testing.

lected during testing at Rocky Flats, manufacturers frequently make design changes or improvements, and the machine on the market may not be the same as the machine that was tested at Rocky Flats.

If the test results are not available from Rocky Flats, or if the model on the market is substantially different from the test machine, the buyer should ask the manufacturer or distributor for test results. The American Wind Energy Association Standard for Testing suggests that a complete test report should include the following information:

A. Test Machine
- model
- serial number of date of manufacture
- any differences between test unit and production units
- description of electrical system and control system

44 YOUR WIND DRIVEN GENERATOR

- B. Installation Details
  - tower type and height
  - load
  - general wiring schematic
  - test site location
  - site layout drawing showing any potential obstructions within 100 meters of the tower and general topography
- C. Instrumentation
  - general schematic
  - name and model number for each piece of equipment plus information on accuracy
  - anemometer location
- D. Miscellaneous
  - data collection methods
  - correction procedures employed
  - final results
- E. Appendix
  - date and time of test
  - general weather conditions
  - wind direction
  - temperature and barometric pressure
  - corrections used

If a manufacturer or distributor is unable to provide test results, the buyer should ask whether the power output (kilowatts versus wind speed) or the energy production (kilowatt-hours versus mean wind speed) is calculated or measured. If the data is calculated, expect that the actual performance will differ substantially. The extent of difference can be seen in the power curve in Figure 4–7 which compares actual test results with a calculated curve.

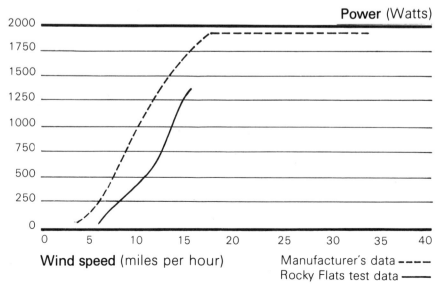

Fig. 4-7. Machine power curve: actual test data vs. calculated performance.

No matter how complete and accurate the testing of a machine, under certain conditions (extreme turbulence for example), the results in operation may differ from test results.

PREDICTING ENERGY PRODUCTION

The key to determining a wind machine's performance, and the objective of most tests, is the *power curve* (Figure 4-8). The curve shows output of the wind machine at a specific wind speed. It also shows various machine characteristics such as rated wind speed, cut-in, and cut-out wind speed.

But the power curve is only part of predicting how much energy a wind machine can produce. The other part is determining what the wind speed will be. It is not enough to know only the annual average wind speed. Because of the cube relationship between the speed of the wind and its energy content, some prediction must be made of the percent of the time the wind occurs at various speeds. There is a statistical formula, *velocity distribution* that predicts what percent of time the wind will blow at various speeds. The *Rayleigh distribution* is a mathematical idealization that is dependent only on the mean wind speed.

This statistical method of determining wind velocity distributions based on the mean wind speed has been shown to agree closely with actual wind data in nearly 90 percent of all cases. By discretely evaluating this function in small increments, 1 mph for example, very little error is introduced over exact integration (Figure 4-9).

The distribution of wind speed can then be integrated with the machine's power curve provided by the manufacturer or distributor. A chart can be prepared that consists of wind speed increments, power output at that increment, and the frequency of occurrence at that increment. For each assumed wind speed,

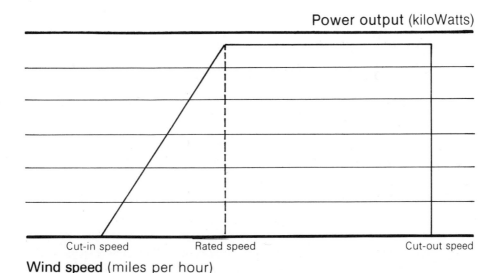

Fig. 4-8. Generic power curve.

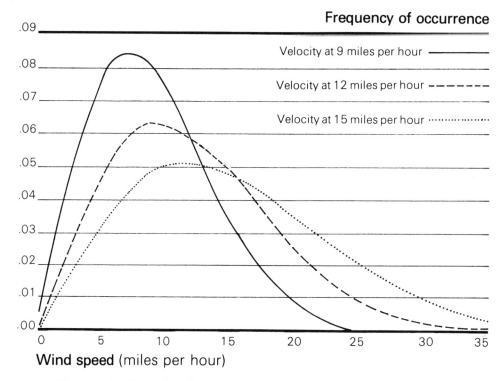

Fig. 4-9. Rayleigh distribution of 9, 12, and 15 mph mean wind speeds.

multiplying the frequency of occurrence by the power produced gives the nominal power at that wind speed. When the products of the frequency times the power ($f \times p$) are summed and multiplied by the amount of time under consideration, the result is the prediction of the energy production. Using the chart on Table 4-1:

at a mean wind speed of 12 mph, the sum is 2.18 kW

if the time under consideration is one year, 8,760 hours:

(2.18 kW)(8,760 hours) = 19,100 kilowatt hours per year

While this process is tedious, particularly if monthly mean wind speeds are used, it will provide a reasonably accurate prediction of the energy that will be produced by a specific machine at a specific site. The largest single cause of dissatisfaction among owners of wind machines is that the machine does not produce as much energy as was expected. While this calculation does not guarantee that there will be "enough" wind, it will predict energy production within the accuracy of the numbers used, and can help avoid unpleasant surprises. Also, it is useful for comparing the energy production of two machines. Generally, this is the technique manufacturers use to predict the energy production of their equipment.

Table 4-1. Integrating Wind Speed Distribution with a Wind Machine's Power Curve.

(v) WIND SPEED (MPH)	WECS POWER (KW) = p	$\bar{v}$ = 9 MPH FREQ(F)	f × P	$\bar{v}$ = 12 MPH FREQ	F × P	$\bar{v}$ = 15 MPH FREQ	F × P
1	0	.0192	0	.0108	0	.0069	0
2	0	.0373	0	.0213	0	.0138	0
3	0	.0533	0	.0311	0	.0203	0
4	0	.0664	0	.0400	0	.0264	0
5	0	.0761	0	.0476	0	.0320	0
6	0	.0821	0	.0538	0	.0369	0
7	0	.0844	0	.0584	0	.0412	0
8	.05	.0834	.0042	.0615	.0031	.0447	.0022
9	.27	.0796	.0215	.0631	.017	.0474	.0128
10	.56	.0735	.0412	.0632	.0353	.0492	.0275
11	.91	.0660	.060	.0620	.0564	.0503	.0457
12	1.3	.0576	.075	.0597	.0776	.0507	.0659
13	1.8	.0490	.0882	.0564	.101	.0503	.0905
14	2.4	.0406	.0974	.0524	.126	.0493	.118
15	3.0	.0328	.0984	.0480	.144	.0477	.143
16	3.7	.0259	.0958	.0432	.160	.0457	.169
17	4.6	.0200	.0920	.0383	.176	.0433	.199
18	5.6	.0151	.0846	.0335	.187	.0405	.227
19	6.7	.0111	.0744	.0289	.194	.0376	.252
20	8.0	.0082	.0656	.0246	.197	.0346	.277
21	8.75	.0056	.0490	.0207	.181	.0314	.275
22	9.2	.0039	.0359	.0171	.157	.0283	.260
23	9.4	.0026	.0244	.0140	.132	.0253	.238
24	9.5	.0017	.0161	.0113	.107	.0224	.213
25	9.5	.0011	.0104	.0090	.0855	.0197	.187
26	9.5	7.2E-4	.0068	.0071	.0674	.0171	.162
27	9.5	4.5E-4	.0043	.0055	.0522	.0148	.141
28	9.5	2.7E-4	.0026	.0042	.0399	.0127	.121
29	9.4	1.6E-4	.0015	.0032	.030	.0107	.100
30	9.4	9.4E-5	8.8E-4	.0024	.0226	.0090	.0846
31	9.4	5.4E-5	5.1E-4	.0018	.0169	.0075	.0705
32	9.3	3.0E-4	2.8E-4	.0013	.0121	.0062	.0577
33	9.3	1.7E-5	1.6E-4	9.5E-4	.0088	.0051	.0474
34	9.2	8.9E-6	8.2E-5	6.8E-4	.0063	.0042	.0386
35	9.0	4.7E-6	4.2E-5	4.7E-4	.0042	.0034	.0306
		.9983		.9980		.9871	
TOTALS			1.051 KW		2.176 KW		3.656 KW

(Note: 1.2E-5 is 1.2×10^{-5})

REVIEWING STANDARDS

Voluntary, industry-wide standards are being developed by the American Wind Energy Association (AWEA), the American Society for Testing and Materials (ASTM), the American Society of Mechanical Engineers (ASME), and the National Fire Protection Association (NFPA).

The AWEA has a broad-based standards program covering all aspects of wind machine design and operation. AWEA and ASTM are cooperating to develop standards to be issued eventually by the American National Standards Institute (ANSI) and National Standards. The standards are being developed with widespread industry and public participation and represent the largest effort to ensure quality wind equipment. Consult the American Wind Energy Association and the manufacturer to determine if the equipment complies with these standards. In some cases, the manufacturer may have a good reason for not complying, but if this is the case, the buyer should know the reason.

The American Society of Mechanical Engineers (ASME) is developing standard performance *test codes*. These test codes are intended for use by utilities purchasing wind systems larger than 100 ft in diameter. They describe procedures by which the utility can test a machine after it is installed to ensure it performs up to the specifications agreed on by the utility and the manufacturer.

The National Fire Protection Association oversees periodic revision of the National Electric Code (NEC). Local communities use the NEC as the standard requirements for electrical safety. NFPA is presently developing a new section for wind energy systems, particularly those connected to the utility grid. The manufacturer or distributor should be able to demonstrate that a specific machine complies with the provisions of the National Electrical Code. If not, it is unlikely a building permit would be issued for the erection of that machine.

TALKING WITH OTHER USERS

After comparing specifications, test results, energy production, and compliance with standards, you will have narrowed your choices considerably. Speak with others who own and use the specific machines under consideration. The manufacturer or distributor should provide the names of two or three individuals who have purchased equipment from them and are willing to talk about their experiences. In addition to asking questions about the equipment, it is a good opportunity to learn about other aspects of the installation:

- What was the system's total installed cost?
- What was the reaction of the utility?
- Was it difficult to obtain a building permit?
- Is the energy production what was expected?
- Is the machine insured under the homeowner's policy?
- What are the maintenance requirements?
- How has the distributor performed? the manufacturer?
- Is any performance data being collected?

This is a small sample of the questions that may be asked. A wind machine owner can be a valuable source of information, not only about the equipment, but about the company, and the range of issues that are raised once the decision to purchase a wind machine is made. If a manufacturer or distributor is not able to provide the names of individuals who have purchased a wind machine from then, you should proceed with caution.

USING CONSULTANTS

Another way to help ensure a sound investment and a good machine is to hire a consultant. This should be done early in the decision-making process, before becoming committed to a specific machine.

Historically, the largest single reason that wind machine owners are dissatisfied has been that the machine doesn't produce as much energy as was expected. Low energy production is almost always due to poor assessment of the available wind. A consultant can help assess existing wind data or take onsite wind measurements. The consultant can also help prepare material to present to the utility, bank, or zoning board.

Fees will vary depending on the amount of time involved in a specific project, and the experience and skills of the firm retained. However, as long as wind energy remains a novel technology, it is advisable to seek expert assistance.

Reading specification sheets, test results, performance predictions, standards, talking to other users, and consultants all will help to ensure that your wind energy investment is a good one.

5
A WIND MACHINE INVESTMENT

A range of financial analysis tools are discussed, including present value, discounted cash flow, internal rate of return, annualized cost, economic utilization factor, and payback.

The largest single factor in the decision to convert to wind energy is economics. Many people are confused by the range of financial terms and tools: life-cycle cost, present value, payback, internal rate of return, and discounted cash flow. Though complicated these tools are necessary to analyze the wind system investment, which requires large initial cash outlays, and returns benefits many years in the future.

At the time of this writing, a typical wind machine costs from $2,000 to $4,000 per kilowatt capacity. Although this is a common reference point, it doesn't describe how much energy the machine will produce. This is a substantial cash outlay, like paying for many years of electric bills at once. The cost of similar wind machines can vary by manufacturer, depending on the number of machines they build per year.

WIND MACHINE COSTS

Before looking at the tools to analyze a specific wind machine investment, consider what comprises the cost of a $22,000 wind machine. The table (p. 51) is an estimate of the manufacturer's cost of producing a wind machine with a rotor approximately 33 ft in diameter, at a rate of 1,000 machines per year, prepared by the Department of Energy.

This example of machine cost is based on a number of assumptions that will not be true in every case, and this outline is, at best, a simplification. Production levels could fluctuate, or a particular site may be more or less costly to install a wind machine. However, the chart does give an idea of how the costs of a wind system accumulate.

The largest single expense to produce a wind machine is the cost of materials, which would not be affected by mass production. The difference in price between a mass-produced machine and today's machines that are being made on a limited scale, is a reduction in the amount of labor required to assemble the machine. Installation costs are a large part of the total cost, but they can vary significantly depending on machine size, design, and site location. For example, some machines have built-in winches and will not require the rental

	HOURS	RATE	AMOUNT
Material			$ 7,136
Material Handling		10%	714
total material			$ 7,950
Engineering Labor	4	$12.50	50
Engineering Overhead		100%	50
Manufacturing Labor	143	$ 7.50	1,073
Manufacturing Overhead		150%	1,610
Other direct Costs			30
FACTORY COST:			$ 10,763
General and Administrative Cost		15%	1,076
TOTAL COST:			$ 11,839
Profit		15%	1,776
SELLING PRICE (FOB factory):			$ 13,615
To this price is added:			
Shipping (250 miles)			500
Distribution cost (25% selling price)			3,403
Installation			5,000
TOTAL INSTALLED COST:			$ 22,518

of heavy equipment. Some machines may be erected close to a manufacturer or distributor and the installation crew will not have to travel far, further reducing the installation costs.

Wind energy equipment is expensive, and the costs add up quickly. Before you can determine whether a wind machine is a good investment, analyze the economics carefully. As attractive as the prospect of energy self-sufficiency may be, some people are better off purchasing their power from the utility company.

LIFE-CYCLE COSTS

Wind machine equipment is expensive, and analyzing the wind machine investment requires careful study and economic tools more frequently used in industrial or commercial investments than in an individual's economic decisions. A key concept is known as *life-cycle cost*. Life-cycle costs are the sum of all benefits and costs incurred over the life of the investment. For example, in the construction of a building—whereas the original cost of a building may be lower without expenditures for energy conservation such as extra insulation, the sum of the original cost of the operating costs over the life of the building may be many times greater than the original savings. Calculating the original cost and the operating costs for alternative investments is the basis of life-cycle costing.

The costs and benefits of owning a wind system occur in complex patterns over an extended period of time. With the life-cycle cost approach, these costs and benefits can be systematically assumed to determine the economic rationale for buying a wind machine or buying electricity from the utility.

To arrive at the life-cycle cost, we must describe the costs and benefits of wind machines and then determine how they interact based on present value, levelized cost, cash flow, and payback. This chapter does not provide the analysis of a specific machine but introduces and defines the tools you need to assess your wind energy investment.

COSTS AND BENEFITS

A typical list of the costs of owning and operating a wind machine includes:

- The wind machine and tower
- Shipping
- Installation
- Wind measurement
- Financing costs
- Insurance
- Property taxes
- Maintenance costs

Various financial benefits are realized from owning a wind system:

- The energy produced
- The energy sold to the utility
- Tax incentives
- Tax deductions

The yearly costs of owning a wind system have to be compared with the benefits, and with the yearly cost of buying power from the utility.

Wind machine costs are immediately reduced by financial incentives from federal state and local governments. A complete list of incentives by state is provided in the Appendix.

PRESENT VALUE

The value of money is not fixed. It shrinks and swells over time. If you leave $100 in a savings account drawing 10 percent interest, it is worth $110 after a year (assuming the interest is paid once per year). Another example is the effect of inflation on money. One hundred dollars kept in your pocket for one year has only $90 in buying power at the end of the year if the inflation rate is 10 percent.

These examples consider the value of money moving forward in time. Given an amount of money in the future, we can determine its present value today. For example, let us take a U.S. savings bond which promises to pay $100 in ten years. How much should you invest to receive that $100? The answer is derived from a formula that determines the present-value discounting. The formula is:

$$PV = \frac{V_n}{(1 + d)^n}$$

where

PV = the present value of the investment
V_n = the value of the capital at the end of n years

A WIND MACHINE INVESTMENT   53

d = the discount rate, which could be the rate of inflation or the interest rate available on an alternative investment.
n = the number of years over which the capital is invested

Let's assume that the U.S. savings bond would pay $100 at the end of ten years, and the general rate of inflation is 10 percent. Then:

$$V_n = \$100$$
$$d = 10\%$$
$$n = 10 \text{ years}$$

and:

$$P = \underline{} = \$38.55$$

In other words, after discounting each of the ten years in the waiting period at a 10% annual compound rate, the present value of the $100 received ten years from now is $38.55. Discounting tables have been calculated and are provided in the Appendix. Simply by selecting values for d and n, you can determine the value for the denominator of the equation.

It is more difficult to analyze the wind system investment because the annual profit or loss varies significantly from year to year. Typically, the value of a wind machine will increase over time, as the cost of the electricity it displaces rises, and the costs of owning the system remain relatively constant. If energy cost and benefit projected to occur during the life of the wind machine were brought back in time to the present, and all these costs and benefits were summed (Σ), we could determine the present value of the investment. Present value of this cash flow, expressed as a mathematical formula is:

$$PV = \sum_{i=1}^{n} \frac{\text{benefits in year } i - \text{costs in year } i}{(1 + d)^i} + \frac{\text{salvage value}}{(1 + d)^n}$$

where

PV = present value of investment
benefits = financial benefits in each specific year
costs = financial costs in each specific year
i = specific year in life of the investment
  = first period in the sum = 1
n = total number of years of the investment
  = last period in the sum
d = discount rate

There are many ways to use this basic equation. The discount rate can be set as the interest rate charged by the bank for a loan to make the initial capital outlay. If cash is paid, the opportunity cost or the interest rate earned by investing the money somewhere else can be used as the discount rate.

54 YOUR WIND DRIVEN GENERATOR

Another more difficult way to use this formula is to fix the present value as the original out-of-pocket expenses, and solve the equation for d. This gives the internal rate of return, or the return on the money invested.

The present value approach is also known as *discounted cash flow*. It is a discounted cash flow because, as in the case of a wind machine investment, the annual expenses and benefits are not uniform and extend over a number of years during which the value of money changes.

THE COST OF UTILITY POWER

To assess the benefits of owning a wind system you must compare cost for your machine with the cost of utility power into the future. To project future utility costs, start by reviewing past electric bills. Figure 5-1 shows the actual rates of utility energy since 1976 experienced by a New England residence. The rate of

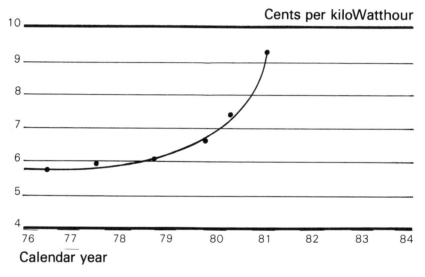

Fig. 5-1. Historical cost per kilowatt-hour of utility electricity.

growth in utility bills is steep and escalating. It is not appropriate to use the utility bill for the year the wind machine is installed to determine the future costs of utility power. Utility costs will certainly rise—the question is how much?

There is no rule for how fast utility prices will rise. The graph shows that utility bills for this New England residence increased 7% in 1978, 12% in 1979 and 30% in 1980. You will have to estimate how fast prices will escalate over the life of your wind system. For example, if we assume the utility bill will increase more rapidly than it did in 1979, but not as rapidly as it did in 1980, an annual increase of 20% is chosen. The estimated cost of utility energy can be determined as follows:

```
year 1:   .095 ¢/kWh  × 1.20 =
year 2:   .114        × 1.20 =
year 3:   .137        × 1.20 =
year 4:   .164        × 1.20 =
year 5:   .197        × 1.20 = .236 ¢/kWh
```

This is a dramatic increase, doubling in just four years. However, this doesn't account for the inflation component of the rising costs. Because the present value approach does not account for the time value of money (essentially inflation), the escalation factor in determining the value of utility energy displaced should be at some rate above the general rate of inflation.

The present value or discounted cash flow method of analyzing an investment can determine whether the investment is a good one (if the present value is greater than the cash outlay or the return on a competing investment), but it does not describe how much better (or worse) an investment in a wind machine would be than continuing to purchase electricity from the utility. For that, we need to look at the cost of energy from the utility compared with the cost of energy over the life of the wind machine. Annual utility payments have to be computed at some escalation rate, summed for the anticipated life of the wind machine, and averaged. Comparing the costs of energy can be done in "constant" dollars, discounting the effects of inflation. This requires that we assume an escalation rate for utility power that is less that the 20% rate in the example above. If we assume that the residence in New England uses 12,000 kWh/yr at .095 kWh, and that rate will increase at 5% per year above inflation, over ten years the average annual utility costs will be:

```
Year 1:  12,000 × .095  =   $1,140  × 1.05 =
Year 2:                      1,197  × 1.05 =
Year 3:                      1,256  × 1.05 =
Year 4:                      1,319  × 1.05 =
Year 5:                      1,385  × 1.05 =
Year 6:                      1,454  × 1.05 =
Year 7:                      1,527  × 1.05 =
Year 8:                      1,604  × 1.05 =
Year 9:                      1,684  × 1.05 =
Year 10:                     1,768  × 1.05 =

TOTAL 10 YEAR COST:       $14,334
AVERAGE ANNUAL COST:    $ 1,433
average ¢/kWh:               .119
```

COST OF ENERGY FROM A WIND MACHINE

A common method of determining the cost of energy (in ¢/kWh) for a wind machine is the *fixed charge rate* method. This method does not reflect the time-value of money, but it is adequate to compare the cost of wind energy to the cost of utility energy. The fixed charge rate method uses a single figure expressed as a percent of the capital cost of the wind machine as the annual charges. It is expressed by the following formula:

$$¢/kWh = \frac{(\text{capital cost})(\text{fixed charge rate})}{\text{annual kilowatt-hour production}}$$

The capital cost includes all the costs of the machine delivered and installed. This can be reduced by any available incentives such as the federal income tax credit. The annual charge rate reflects interest, depreciation, taxes, operation and maintenance costs. Although this varies considerably from situation to situation, a typical annual charge rate is 18%. The annual kilowatt-hours production for a specific site and a specific machine can be estimated as shown in the previous chapter.

56 YOUR WIND DRIVEN GENERATOR

For example, if we assume a wind machine would produce 45,000 kWh/yr at a specific site, and could be delivered and installed for $20,000, the cost of energy would be:

$$\frac{(\$20,000)(.18)}{45,000} = 8¢/kWh$$

The fixed charge rate is a crude method for determining the cost of energy, but it is useful to compare the costs of various alternatives.

ECONOMIC UTILIZATION FACTOR

The tools provided above enable an individual to determine whether wind generated electricity will be cheaper than utility generated power. However, because sometimes you will sell power to the utility and other times buy power, we must examine the *economic utilization* factor.

The economic utilization factor assumes that a certain amount of the energy generated at a specific site, in excess of demand, will be sold to the utility. This is represented by the following formula:

$$\text{Economic Utilization Factor} = (\text{direct use}) + (1 - \text{direct use})(\text{utility purchase rate})$$

where

$$\text{direct use} = \text{the energy consumed directly expressed as a percent}$$
$$\text{utility purchase rate} = \text{utility buy-back price as a percent of retail price}$$

For example, if 50% of the energy generated is used directly onsite, maximizing the utility buy-back rate will yield the largest economic utilization factor.

All these factors can be presented in a graph form. Figure 5-2 shows the effect of various economic utilization factors (0.6, 0.75, 0.9), and various average wind speeds from 12 mph to 16 mph, assuming a wind machine that costs $20,817 installed and generates 31,000 kWh/yr in a 14 mph annual mean wind.

All the characteristics of a specific wind machine and installation can be integrated into a *sensitivity analysis* which shows how specific variables impact the *base case* (Figure 5-3). It is apparent that decreasing the installed cost has approximately the same impact as a better site. Increasing the life of the machine has very little impact, although reducing machine life is as important as other variables.

CASH FLOW AND PAYBACK

Everyone makes decisions based on cash flow projections. Few people are able to purchase the home they want outright. They may buy a home on borrowed money knowing that as time goes by, inflation will soften the impact of the monthly mortgage, and the value of the home will increase. However, no matter how smart the purchase of real estate is in the long run, our ability to meet the monthly payments at the time of the purchase will determine if, and to what extent, we can participate in the housing market.

The purchase of a wind machine involves substantial expenditures early in the life of the machine. The

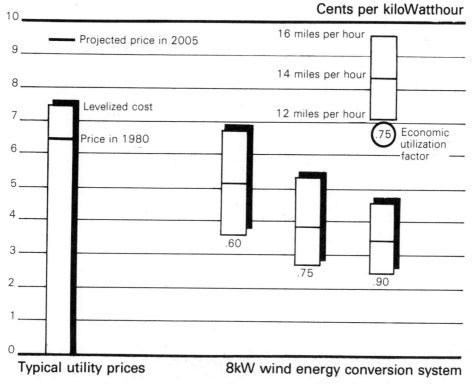

Fig. 5-2. Cost of wind-generated electricity at various economic utilization factors and mean wind speeds.

major financial benefit of ownership, energy production, offsets the early costs later in the life of the machine. A potential wind machine owner may find that the extra expense early in the machine's life prohibits the purchase, regardless of the long-term costs savings.

Life-cycle cost analysis has to be supplemented with a year-by-year cash flow analysis. This type of analysis identifies the combinations of installed costs, interest rates, competing energy costs, and other factors that would result in a large initial negative cash flow.

Even if the wind machine purchaser is convinced that the cost of energy of the wind machine is less than the cost of purchasing utility power, or that buying a wind machine is a better investment than other alternatives, he or she may not be willing to purchase a machine if the total investment in the machine is not returned in a relatively short time. This is known as *payback*. Many socialist countries that do not recognize the existence of interest (because according to their theories, all value springs from labor, and capital, is dormant), use payback to evaluate the relative attractiveness of projects. While most economists have abandoned payback as a tool of economic analysis, it is common enough that it should be understood.

There are two points in the payback cycle that are important to note. The first point is the year when the wind machine owner pays no more out-of-pocket for the wind-produced electricity than he would pay for the utility power. The second point is the year when the wind machine owner has recovered all the extra expenses invested in the wind machine. This is the *cumulative breakeven* year.

58 YOUR WIND DRIVEN GENERATOR

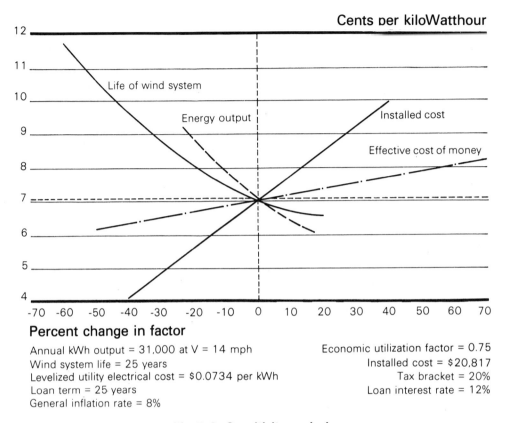

Fig. 5-3. Sensitivity analysis.

The chart on Table 5-1 is an example of a cash flow analysis and shows the annual breakeven year and the cumulative breakeven year. A cash flow analysis of a wind machine investment will be very different in each situation because of the large number of variables. The assumptions made in the example include:

Total installed cost:	$20,000
Down payment:	33% or $6,600
Loan term:	7 years
Interest rate:	19%
Maintenance costs:	2.5% of installed cost: $550/year
Energy production:	55,000 kWh/yr, 75% consumed directly (41,250) displacing 7¢/kWh electricity; and 25% sold to the utility (13,750) at 5¢/kWh
Utility escalation:	3%/yr above inflation
Tax credits:	$4,000 federal and $1,000 state

This analysis does not include the tax advantages of this investment. For example, an individual in the 30% marginal tax bracket would, by deducting the interest expense, break even in the second year.

Table 5-1. Wind Machine Cash Flow Analysis.

YEAR	0	2	3	4	5	6	7	8	9	10
Down payment	$6,600									
Principal payment	1,070	1,273	1,515	1,803	2,145	2,553	3,038	0	0	0
Interest	2,546	2,342	2,100	1,812	1,470	1,062	577	0	0	0
Operation & maintenance	500	500	500	500	500	500	500	500	500	500
Insurance	50	50	50	50	50	50	50	50	50	50
Property tax	65	65	65	65	65	65	65	65	65	65
TOTAL COSTS:	10,831	4,230	4,230	4,230	4,230	4,230	4,230	615	615	615
Energy displaced (@7¢/kWh + 3%/hr)	2,887	2,974	3,063	3,155	3,249	3,347	3,447	3,551	3,657	3,767
Energy sold (@5¢/kWh + 3%/yr)	687	707	728	750	773	796	870	844	870	896
TOTAL INCOME:	3,574	3,681	3,791	3,905	4,022	4,143	4,317	4,395	4,527	4,663
Federal tax credit	4,000									
State (Vermont) credit	1,000									
TOTAL CREDITS	5,000									
ANNUAL PROFIT (LOSS)	(2,257)	(549)	(439)	(325)	(208)	(87)	87	3,780	3,912	4,048
CUMMULATIVE PROFIT (LOSS)		(2,806)	(3,245)	(3,570)	(3,778)	(3,865)	(3,778)	2	3,914	7,962

Annual breakeven: Year 7
Cumulative breakeven: Year 8

Also, because business tax credits and tax treatment are substantially different from residential ones, the cash flow and annual profit or loss can change substantially. The tax implications of a wind machine investment are complex, and potential purchasers are advised to consult an accountant or tax attorney to determine exactly what benefits are available in their specific situation.

6
THE SMALL WIND SYSTEMS AND THE UTILITY

The Public Utility Regulatory Policies Act (PURPA), utility buy-back rates, technical concerns, and approaching the utility

Until wind machines are well established in America, wind producers should anticipate lengthy negotiations with the local utility before interconnection. A well-prepared and responsive wind system owner can move quickly to resolve the issues raised by the utility. It is important to understand the utility's concerns and the context within which it operates.

ELECTRICITY IN AMERICA

Historian Alistair Cooke, in his television series "America" identified the windmill as the single most important use of technology in the settlement of the West. By using the wind to pump water in the late 1800s, isolated land became inhabitable for homesteads and ranches. By the early 1900s, thousands of farmers used wind energy to generate electricity. These farms were far from the cities and towns that already had electrical distribution systems and were far apart from each other. It was not a profitable proposition to lay hundreds of miles of wire to deliver electricity to a few customers who wouldn't use much anyway. The electrification of rural America, in the words of a utility executive at that time, was a "utopian dream."

The Rural Electrification Administration (REA) changed that by offering low-cost loans for the construction of distribution systems serving remote farms. Private utilities still did not take advantage of the incentives, and communities which did not have electrical power formed rural electric co-operatives (REC) to take advantage of the REA subsidies. Today, more than 1,000 rural electric co-operatives serve over 25 million customers in 46 states. Most are distribution and transmission utilities that purchase power from other utilities. RECs own and maintain nearly 44% of the distribution lines in the United States, but only about 2% of the generating capacity. The large investor-owned utilities are the primary generators of electrical power.

Today, potential residential, agricultural, commercial, and industrial users of wind energy have commercial utility electricity available. Therefore, most new wind machines are intended to operate parallel with the utility distribution grid. This operation, where consumers of electricity also become producers, and the utility becomes a "consumer" signals a quiet revolution in the electric utility industry.

In twenty years, the utility industry will probably bear faint resemblance to the utilities of 1982. The distinction between energy producers and consumers, once clear, is now clouded. Wind energy is not the only technology supplying power to the utilities. Many industries today cogenerate energy, simultaneously producing heat and electricity. The heat is used in industrial processes, and the electricity reduces the electrical demand from the utility, or is fed back into the utility grid. In the future, photovoltaic cells on the roofs of homes will feed excess power into the utility grid. Wind energy is presently the most economically competitive source of electricity from the new, renewable sources of energy, and many of the problems associated with the changing relationship between the utility and their customers will be experienced by wind users.

Utilities perhaps more than any other industry have fallen victim to the uncertainties and rapid changes in energy economics. Growth in demand has failed to follow historic trends—the cost of new plant construction and the cost of fuel have risen precipitously, and the time required to obtain licenses and construct new plants has risen unpredictably. Many utilities are overextended with major construction projects, inaccurate demand projections, soaring interest rates, and construction delays that double or even triple the projected cost of new construction. These uncertainties have made planning much more difficult at a time when the cost of an error has greatly increased. Regulatory restrictions and other factors have contributed to the problem, often making it difficult for utilities to maintain acceptable net income. Utility stocks have fallen below book value; and bond ratings, nurtured for decades, have eroded. Orders from new electric plants have declined dramatically as the industry and its investors reevaluate their options.

While the origins of the problem are complex, a major cause has been a consistent underestimation of both the cost of energy from new plants and the speed with which the users reacted to higher prices and the anticipation of higher prices. Declining demand is reinforced by new technologies for increasing efficiency and by onsite power generation such as wind machines, for example. New technology, new economics, and a new political environment have combined to create a fundamental shift in the role of the utility. As is often the case with major societal changes, a recent law reflects this change.

THE PUBLIC UTILITY REGULATORY POLICIES ACT (PURPA)

PURPA became law as part of the National Energy Act of November 8, 1978. The broad purpose of the law is to encourage the conservation of energy and the efficient use of energy resources by utilities. One way it does this is to encourage cogeneration and small power production. Small power production is defined as facilities up to 80 megawatts (MW) (a megawatt equals 1,000 kilowatts) that produce electricity using primarily biomass, waste, or renewable resources such as solar, wind or hydro power. Sections 201 and 210 of PURPA encourage these forms of power production by mandating that utilities purchase and sell power to these facilities and exempting such facilities from federal and state utility regulation. All electric utilities, regardless of size, ownership, or whether or not they have generating capacity, must purchase power from such facilities for a reasonable fee.

In February and March of 1980, the Federal Energy Regulatory Commission (FERC) issued regulations implementing Section 201 and 210 of the law. Although the regulations are specific in many areas, state utility commissions have a great deal of flexibility implementing them. Implementing PURPA will be a challenge, both because the law breaks new ground and because some of the new technologies are not well understood by the utilities or the regulators.

Utilities and state regulators are neither accustomed to consumers of electricity becoming producers, nor to utilities becoming consumers. Most utilities are not used to calculating the cost price, and even fewer are familiar with the refinements of marginal cost pricing required under PURPA. While many utilities and commissions are experienced with cogeneration in large industrial plants, they are not familiar with wind energy.

This discussion of PURPA cannot provide a comprehensive answer to the complexities of the law—they are still being worked out. Utilities, state regulators, and small power producers need time and experience to find the best ways to implement this law which has been called the "Magna Carta" for small energy producers.

Before PURPA, individuals who wanted to generate their own power in connection with the utility grid faced three problems. Utilities sometimes refused to buy their excess electricity or they paid a very low rate. Utilities sometimes refused to sell power to an independent producer, or did so only at a very high rate. And finally, the small power producer selling electricity was subjected to cumbersome regulation as utilities. PURPA addresses each problem:

- Utilities must offer to buy energy and capacity from small power producers at the marginal rate the utility would pay to produce the same energy.
- Utilities must sell power to these facilities at non-discriminatory rates. Further, qualifying facilities are entitled to simultaneous purchase and sale. They have the right to sell all their energy produced to the utility and purchase from the utility all energy needed for consumption.
- Small power producers are exempt from most federal and state financial, rate, and organizational regulations that apply to utilities, except PURPA.

There are some things PURPA does not do:

- PURPA does not prevent an electric utility and a small power producer from making an agreement or purchase even if the terms are different from those otherwise required under PURPA.
- PURPA does not affect the validity of any existing contract for the purchase of power. Thus, the PURPA approach to determining purchase rates are applicable only when the small power producer and the utility cannot agree on a rate themselves.
- PURPA does not preempt states from legislating rates higher than the utility's avoided cost, although it does forbid lower purchase rates for electricity from a small power producer's new capacity.
- PURPA does not set a minimum size for small power producers.

UTILITY BUY-BACK RATES

The issues associated with the implementation of PURPA are complex and many of the provisions are being contested in the courts. Perhaps the most interesting and controversial is the question of determining the buy-back rate. The buy-back rate at which the utility purchases energy or capacity from the small power producer is to be established by the public utility regulatory body in each state. This rate is determined by the utility's avoided cost.

FERC defines *avoided cost* as the incremental or marginal cost to an electric utility of energy or capacity, or both, which the utility would have to generate itself or purchase from another source if it did not buy power from the small power producer. Several important features should be considered:

1. If the power from the small power producer is sufficiently reliable, the purchase rate should reflect capacity as well as energy credits.

2. The avoided cost includes not only present, but also future costs.
3. The avoided cost is not the utility's cost of service, nor is it the average system's cost, nor is it the small power producer's cost of production.

To figure the utility's avoided cost, the public utility commission begins with the utility's system cost data. However, many other factors are taken into account "to the extent practicable," according to PURPA. They include:

1. The ability of the utility to dispatch the facility.
2. The reliability of the facility either expected or demonstrated.
3. The terms of the contract including duration of the obligation, termination notice requirements, and sanctions for non-compliance.
4. How well the small power producer schedules outages coordinated with the utility's period of low demand, to enable the utility to avoid certain costs.
5. The usefulness of both energy and capacity of the small power producer during utility system emergencies.
6. The individual and aggregate value of energy and capacity form the small power producers in the utility system.
7. The small capacity increments and shorter lead times necessary for additions of capacity from small power producers. This allows the utility to accommodate changes in forecasts of peak demand which can result in cost savings.
8. The extent to which the energy or capacity from the small power allows utilities to defer expansion and reduce the level of fossil fuel use. A utility with extensive capacity expansion plans is likely to benefit more from the energy and capacity provided by small power producers than is the utility with excess capacity.
9. The extent to which line losses are reduced when the electricity supplied by the small power producer is closer to the point of use than the utilities' generators.

Other factors may also determine the buy-back rate. For example, a utility is not required to purchase power from a small power producer when the utility's avoided cost is less than zero. However, a commission could figure such a period into the rates instead of attempting to change the purchase rate from moment to moment. Also, in the future there may be a standard purchase rate for facilities under 100 kW in capacity. The purchase rate may be different for wind systems than for a small hydroelectric installation.

Despite the complexity of the law, the procedures a utility would use for calculating a utility's net avoided cost, and thus its rate for the purchase of energy from small power producers, is quite straightforward. It consists of the following steps, and is applicable to any renewable energy technology.

1) Develop the best capacity expansion plan for the utility, without the purchase of energy from small power producers. This plan will be based on data FERC requires from the utilities.

2) Estimate the performance of your wind energy system, including hourly and monthly variations in wind speed at the machine's site and in the utility's service area. This would determine the hourly energy production from the wind systems. From the performance and load data, calculate the effective carrying capacity and energy displacement of the wind systems. Subtract the output of the wind systems from the utility's base load data.

64 YOUR WIND DRIVEN GENERATOR

3) Develop a new optimal capacity expansion plan based on the reduced load data. Then, calculate the difference in the present value of the cost between the two expansion plans. This difference is the utility's net avoided cost as a result of purchasing power from the wind system.

4) Convert the avoided cost into a rate. One way of calculating the rate would be to set it equal to the annuity whose present value is equal to the difference in the present value costs of the two expansion plans.

The actual application of these steps raises two problems with wind systems, both in the second step. It is difficult to determine wind-generating capacity in the utility's service area in terms of power and number of wind energy systems. Few utilities have developed estimates. In addition, data about wind speed is not complete and the performance characteristics of future wind machines are not available.

This is not the only way to calculate a purchase rate. For example, the state of New Hampshire does not use this approach. The New Hampshire buy-back rate, and the methods by which it was determined is shown on the chart on Table 6–1.

Table 6–1. New Hampshire Buy-back Rates.

ITEM	AVOIDED COST	DESCRIPTION
ENERGY ITEMS		
Base Fuel Costs	61.81 mills/kWh	Based on oil at $35/barrel and heat rates at base plant (largest utility's most recently constructed and most efficient oil generating station)
Adder for daily peak	6.18	Needed because of times when base plant is either not on line or not following system load.
Correction for forced outage	4.33	Utility must substitute less economical units than base plant.
Inventory cost	1.89	
Operation and maintenance	2.10	
TOTAL ENERGY AVOIDED COST:	76.31 mills/kWh rounded to 77 mills/kWh	
CAPACITY	5.00 mills/kWh	Based on capacity from New England Power Pool at $22/kW.
Total Energy and Capacity Rates:	81.31 mills/kWh rounded to 82 mills/kWh	

1 mill = .1¢

Today, the rate for selling your energy to the utility will usually be determined by negotiations between the utility and the individual. These discussions will proceed more smoothly if the wind system owner understands the utility's concerns (Table 6–2). Therefore you should understand basic technical issues associated with operating a wind machine in parallel with the utility grid.

UTILITY CONCERNS

Electric utilities must provide high quality, reliable electricity to all customers at a fair price. A customer who wants to interconnect his own generating system with the utility's must demonstrate that it will not affect the quality or reliability of electrical service to other customers.

While PURPA provides the small power producer with certain rights regarding interconnection, the utility has the right to resolve many of the technical issues. Their attitude regarding small power producers

THE SMALL WIND SYSTEMS AND THE UTILITY 65

Table 6-2. State Utility Characteristics and Fuel Costs.

STATE OR POOL	(1980 $) FUEL COST[1]			NET GENERATION	
	COAL (¢/10⁶ BTU)	OIL (¢/10⁶ BTU)	GAS (¢/10⁶ BTU)	OIL %	GAS %
N.E. POOL (MA, VT, CT, NH, ME, RI)	178.8	379.5	403.0	55.4	0.9
NEW YORK	146.4	402.8	241.3	38.0	6.1
PJM (PA, NJ MD, D.C.)	131.2	480.9	282.1	17.0	1.9
ILLINOIS	158.3	499.8	283.1	7.6	2.5
INDIANA	125.7	602.7	241.2	1.0	0.4
MICHIGAN	152.4	398.3	263.5	10.7	2.8
OHIO	142.3	504.5	293.6	1.6	.3
WISCONSIN	136.8	514.1	272.4	1.5	3.5
IOWA	128.8	607.7	238.7	1.7	2.8
KANSAS	101.8	460.4	166.4	5.7	1.7
MINNESOTA	98.5	498.9	197.6	1.8	1.7
MISSOURI	116.1	356.7	186.7	1.2	3.3
NEBRASKA	117.3	369.8	168.8	2.3	6.2
NORTH DAKOTA	55.0	605.5	0.0	0.7	0.1
SOUTH DAKOTA	65.8	649.0	196.2	0.6	0.1
DELAWARE	159.1	411.3	349.1	59.4	6.9
FLORIDA	169.8	347.1	134.5	47.9	16.0
GEORGIA	144.3	574.1	287.0	2.5	0.4
NORTH CAROLINA	152.5	583.7	340.8	0.3	0.1
SOUTH CAROLINA	153.6	481.9	257.0	5.6	1.0
VIRGINIA	170.9	408.9	334.0	39.5	0.7
WEST VIRGINIA	133.6	622.2	270.5	0.6	—
ALABAMA	158.7	626.0	250.7	.7	.9
KENTUCKY	128.4	627.9	210.2	.2	.1
MISSISSIPPI	172.6	286.0	194.2	37.7	30.6
TENNESSEE	152.5	653.8	—	.5	—
ARKANSAS	121.4	340.5	211.4	27.0	15.5
LOUISIANA	187.2	342.9	157.3	17.7	82.3
OKLAHOMA	123.5	—	166.1	0.1	82.9
TEXAS	113.6	439.1	173.7	1.4	73.2
ARIZONA	96.1	464.9	245.5	8.2	13.5
COLORADO	81.5	477.7	244.8	1.6	12.5
MONTANA	43.1	477.7	387.2	0.4	1.2
NEVADA	125.7	387.0	278.7	5.9	26.0
NEW MEXICO	50.2	523.6	242.1	1.9	26.3
UTAH	111.7	595.1	195.3	1.1	4.2
WYOMING	52.5	773.4	262.8	0.3	0.3
CALIFORNIA	111.7	480.2	326.2	40.6	28.8
NWPP (WA., ORE., IDA.)	102.1	368.5	362.6	0.6	0.3
ALASKA	N/A	N/A	N/A	13.6	60.9
HAWAII	—	364.0	—	99.7	—

1) Fuel costs present for March 1980 in (1980 $).

varies widely. Some utilities view wind systems as conservation devices because they reduce the demand for conventional energy. The wind system is considered a *negative load,* the same as "turning off" an electrical load, such as a sauna or welder.

When the utility considers the wind system a mini-utility, the issues and concerns raised go well beyond the concept of negative load. These concerns have been raised by the large investor-owned utilities and the small rural electric co-operatives. The concerns generally fall into two categories: safety and power quality. Sometimes the concerns are legitimate based on the utility's experience with independent generation and their concern for providing reliable, high-quality power to all users.

Safety factors

Utilities are seeking to avoid events or conditions which are, or may become, a hazard to the public or their personnel. Safety considerations for wind systems connected to the utility include:

- system grounding and shock hazards
- energizing dead utility lines
- lightning

System Grounding and Shock Hazards. System grounding and shock hazards pose a clear threat to utility personnel and public safety. However, in terms of utility interfacing, a primary concern of the utility is that reasonable measures be taken to protect linemen and meter readers. Compliance with recognized electrical safety codes and standards and local building codes are generally sufficient. Article 250 of the National Electrical Code, Grounding, outlines reasonable measures to assure the safety of the general public and the wind machine owner.

Energizing a Dead Utility Line. De-energized utility lines represent the greatest safety threat for utility personnel, particularly energizing an "open" utility line from the customer's end. A radial distribution feeder is usually removed from service by opening a switch at the utility substation. If the wind generator was allowed to continue operation, the utility line might remain energized and pose a potential problem for utility line crews.

Today, all interconnected wind systems are designed to stop generating when the utility line is dead. Some wind machines, such as the induction generator and line-commutated inverter are voltage dependent—they need utility power to operate. Others, such as the self-commutated inverter and the synchronous generator employ switches that open when the utility line loses voltage. Either way, the wind system automatically disconnects when there is no utility line power.

Besides automatic safety features built into the wind equipment, more utilities require a visible, accessible "disconnect" switch to isolate all possible sources of electricity on the distribution line. This is called an *air break disconnect,* a device which is opened by the operator and locked to prevent accidental closure. Disconnects are usually controlled by a dispatcher or line superintendent. If the device is opened electrically, the mechanical link between the operator and the switch is manually removed to prevent accidental closure. To insure that a device is not closed accidentally, a card or tag is hung on the mechanism to prevent closure until the person who has requested the tag has assured the load dispatcher or line superintendent that no personnel are in contact with the equipment and the problem has been solved.

THE SMALL WIND SYSTEMS AND THE UTILITY 67

The combination of automatic features of wind machines and the utility safety procedures ensure safe operation of an interconnected small wind system.

Lightning. Utility distribution systems are protected against induced and direct lightning surges by the use of *surge arrestors* on primary feeders. The utility is responsible for damage caused by a lightning strike to its system. The wind generator must not increase the probability of surges in the utility system. There is a certain probability of a direct lightning strike to all outdoor equipment. A direct strike to the rotor or tower must not result in a surge into the utility's distribution system so, utilities may require surge arrestors between the wind system and the distribution line. This is a common piece of equipment that can be provided by the wind system distributor.

The common safety concerns raised by utilities can generally be addressed to the utility's satisfaction through a combination of present utility work procedures and equipment such as surge arrestors.

Quality of Power

All utilities are concerned about the quality of electricity fed from the wind machine into their distribution system. The term *quality of power* narrowly refers to the harmonic content of the power generated by the wind machine, although the term is often applied as a catchall for the various factors that describe electrical power, such as frequency, voltage control, and power factor. All utilities agree that power quality is desirable but few will define it with engineeering specifications. Although electrical power-generating systems have existed for more than a century, the availability of electricity from customer-owned power systems raises fundamental questions that most utilities have not considered.

The problem is complex because there already exist a broad range of devices, appliances, and loads on the utility grid that affect the quality of power. For example, the use of regenerative drives has been common for a number of years. Most tall buildings use regenerative drives in their elevator systems. The motor that lifts the elevator operates on direct current and can also be driven as a generator. The motor lifts the elevator car. When the car is descending, however, the motor works as a generator, extracting the energy from the elevator as it comes to each stop on the way down. The energy is converted to alternating current and fed in to the building's electrical system. With this device, the energy required to operate the elevator can be reduced by as much as 80%. Many wind machines employ similar technology, except of course, the wind machine's generator is not used as a motor.

A few wind systems scattered through a utility's service area will have virtually no impact on the utility's power quality. However, if 50% of the homes on a distribution line have electrical generators feeding power into the grid, the problems associated with quality of power become critical.

Although there is no standard approach to defining power quality, it is a major concern of the utility and the potential wind machine owner should be familiar with their three major power quality concerns: power factor, harmonics, and voltage control.

Power Factor. In alternating current electricity, both the voltage and the current appear as sine wave. In the United States the frequency is 60 hertz (Hz) which means the waves complete one cycle sixty times per second. The term *power factor* describes the position of the voltage wave form relative to the current wave. By definition, power factor is the cosine of the angle between the voltage and current sinusoids. When they are both in line, that is when they peak at the same point in time, the power factor is 1 (Figure 6-1). When

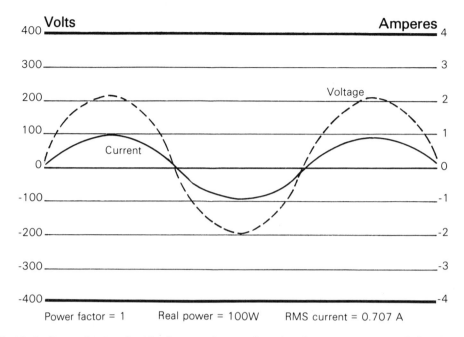

Fig. 6-1. Power factor. An ideal case where voltage and current are exactly in phase.

they are out of line, the power factor is less than one (Figure 6-2). An electrical load can have either a leading or a lagging power factor, depending on whether the current wave form "leads" or "lags" the voltage wave form. Most utilities operate with a slightly lagging power factor.

The problem presented by a lagging power factor is that only the portion of the volts and amps that coincide properly can do real work. This is known as *active power*. To the degree the current and the voltage are out of line, additional current has to be fed into the line. Almost all residential utility customers have slightly lagging power factor due to the inductance of the load. The utility must supply real or active power (kilowatts) as well as the additional power that does not perform useful work, known as *reactive power* (kilovolt-amps reactive or *vars*). This is done by overexcitation of the utility generators, or at the local level by the use of shunt capacitors. Capacitors help supply voltage by injecting reactive power into the system.

An excessively lagging power factor can be a problem for utilities in voltage control as well as sizing of equipment such as transformers, circuit breakers and conductors. The utility's concern in reactive power consumption is for the customer to pay a fair price for the electrical service and not cause variations of voltage service limits. To this end, utilities place power factor limits on commercial and industrial customers. A few do the same for residential customers, but most do not because it is rarely a problem in typical residential areas. To ensure acceptance by the utility, the power factor of your wind machine's electricity should be controlled to between 0.8 leading and 0.9 lagging under all normal operating conditions.

Harmonics. Voltage and current in an ideal alternating current power system are sinusoid wave forms. A certain level of deviation from the ideal case is inevitable in practical systems. These deviations take the form *harmonics*. Harmonics are defined as whole-number multiples of a single frequency that are present in any wave which is not a pure sine wave. In a power system whose fundamental frequency is 60 Hz, there could be

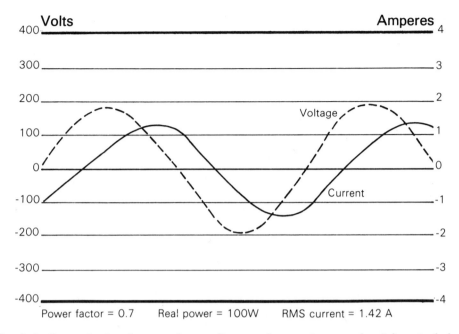

Fig. 6-2. Power factor. A case where voltage and current are moderately out of phase.

smaller wave forms of 180 Hz (3rd harmonic), 300 Hz (5th harmonic), 420 Hz (7th harmonic) and so forth. The fundamental frequency and the harmonics create a composite wave form that is distorted from a pure sine wave as a function of the amount of harmonics present.

The adverse effects of harmonic distortion are strongly system dependent and somewhat unknown. Modern electronic equipment and home appliances are often designed assuming no distortion. The effects of distortion on such equipment is not clear. The capacitor, sometimes used by utilities for power factor control, is sensitive to harmonic distortion. Interference with telephone, radio, and television is possible at high distortion levels.

Most utilities do not have officially acceptable levels of harmonic distortion. The usual criteria is operating experience without failure or customer complaint. If there is no failure or complaint, the utility is likely to accept any level of distortion. In private conversations, some utility executives define the acceptable level of harmonic distortion as not more than 5% Total Harmonic Distortion (THD) and not less than 3% for any one harmonic. If a particular technology or wind machine exceeds these levels, most common harmonics can be attenuated by filters.

Voltage Control. Most utilities are required by regulatory commissions to comply with a national standard for voltage control in providing electrical service within certain voltage ranges to customers. The standard, "Voltage Ratings for Electrical Power Systems and Equipment" (ANSI C84, 1-1977), defines an acceptable voltage range of ±5% for 120/140 volt services. In emergency conditions, they are permitted to operate at ±10%. Utility customers with generation capacity should comply with similar standards.

Voltage is controlled on the utility distribution system by several means. As loads change over the years, voltage regulation equipment must also be changed. New installations are often designed to operate initially without voltage regulators. The most common techniques of voltage control are the installation of capacitors and substation or line voltage regulators. These devices are controlled by timers, manual switches, or power factor relays. Whatever device is used, the purpose is always the same—to maintain service voltage within certain limits.

The wind machines and their interface must not cause service voltages outside acceptable limits. Utilities will probably not have serious voltage regulation problems as a result of small wind machines, because wind machine installations will not occur suddenly. Rather machines will be installed in small capacities throughout the utility service area, and if by chance many are added to a particular feeder, the voltage regulation for normal load growth, and wind machines added to a feeder will influence this normal adjustment only slightly if at all.

Small wind systems installed on another customer's circuit may cause annoying light flicker. If this customer also shares a secondary circuit with other customers, this voltage flicker could cause a problem for the utility. The existence of a voltage flicker problem will depend primarily on the start-up characteristics of the wind machine. If the generator draws a large inrush of current during start-up, voltage drop detectable as a flicker may result. Although no physical damage is likely to result from voltage flicker, complaints from other customers connected to the secondary circuit make this unacceptable.

To avoid voltage flicker, utilities may require the owner or manufacturer of the wind machine to install a starter that limits the inrush of current. If the wind machine owner and the utility agree, isolating the wind machine on the distribution transformer may provide the least-cost solution that allows only the wind machine owner to experience dim lights during generator start-up.

INFORMING THE UTILITY

Each potential wind machine owner should make available to their utility and state or local authorities such as the public utility commission, building inspector, or zoning board enough information for them to assess the safety and reliability of a potential wind machine installation. This could include the following information:

1. A schematic diagram of the wind machine's electrical system similar to the diagrams in Figures 6-3, 6-4, and 6-5.
2. A copy of the wind machine's operating manual which discusses the normal and failure mode operation of the wind machine's electrical system.
3. A table of the wind machine's electrical characteristics including:
   - Type of interconnection hardware, i.e., industrial generator, synchronous alternator, inverter.
   - Rated and maximum alternating current output in kilowatts and kilkovolt-amperes.
   - Rated AC line voltage and allowable variations.
   - Rated AC line current.
   - Power factor or VAR characteristics at 25%, 50%, 75%, 100% of rated power output.
   - Voltage/current harmonic characteristics at 25%, 50%, 75%, and 100% of rated power output.
   - Maximum inrush currents drawn during start-up or normal operation if applicable.

The wind machine manufacturer should provide the diagrams, discussion of operation, and electrical system characteristics to the potential wind machine owner through the distributor. The installer or owner of the wind machine should provide site-specific information and drawings.

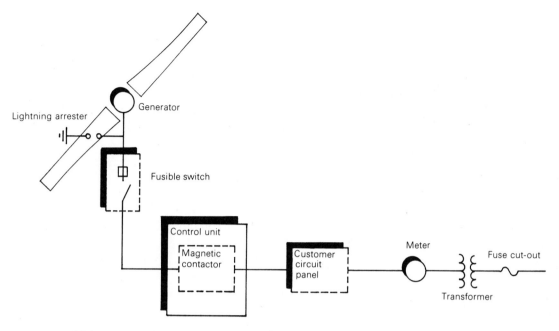

Fig. 6-3. Schematic of wind machine/utility interface: induction generator.

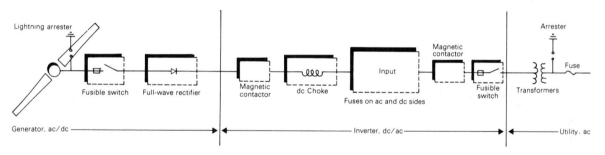

Fig. 6-4. Schematic of wind machine/utility interface: line-commutated inverter.

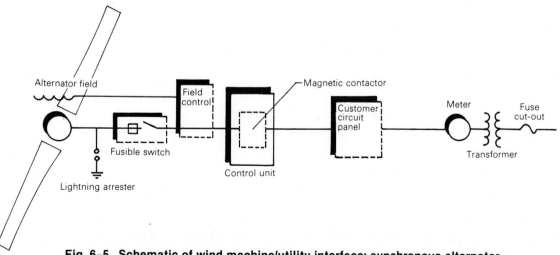

Fig. 6-5. Schematic of wind machine/utility interface: synchronous alternator.

72 YOUR WIND DRIVEN GENERATOR

A complete description of the wind machine's electrical system characteristics will be valuable to everyone involved in the project. Manufacturers should demonstrate that they understand how their machine affects the utility's electricity with respect to safety, reliability, and quality of service. The potential wind machine owner should compare electrical system characteristics, diagrams, and functional characteristics of different machines. The utility needs the information outlined above to accurately assess the safety and quality of service a wind machine will have on their system.

OTHER ISSUES

There are a number of additional utility-related issues which should be considered by the potential wind machine owner. For example, who should bear the costs for interconnection, including the extra meters, surge arrestors, power factor correction, harmonic filters, or an air-break disconnect. If the utility could prove that the equipment was necessary for safe, reliable, quality service to other customers, perhaps the wind machine owner should pay the costs. If the equipment was not necessary, perhaps the utility should pay them. The burden of proof will fall on the wind machine owner, and for that reason, the potential owner should be familiar with utility's concerns. A utility may ask the user to pay all additional costs, even those that are unnecessary. Utilities might install additional metering and test equipment in the interest of determining the effects of the wind machine on their system. Eventually, the state regulatory commission will have to decide who should pay for these costs. The wind machine owner should not automatically agree to all the utility's proposed interconnection costs, but should understand the basis for the proposals and negotiate a fair compromise. In any event, the terms of payment can vary. For example, rather than applying all additional costs at the beginning, payments could be paid on each month's bill.

Liability for damage is another concern of utilities. Rural electric cooperatives have proposed that wind machine owners should buy as much as $1,000,000 in liability coverage for their machines to cover damages to other customers of the utility system. The potential wind machine owner may have to deal with this question in some utility service areas. Though it may be difficult, the utility should prove that there is a legitimate risk to it or to other customers. The wind machine owner should ask the utility to be insured against damage to the wind machine as a result of utility operations. At this time, the whole question is unsettled. Resolving this and other unforeseen complications will ultimately depend on the mutual understanding and cooperation of all parties.

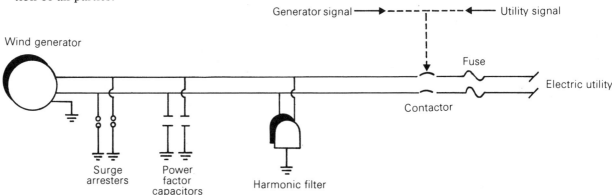

Fig. 6-6. Potential utility power quality requirements schematic.

SUMMARY

Safety and power quality concerns actually cover a broad range of issues about which, in many cases, little is known (Figure 6-6). Research indicates that the problems associated with safety and particularly power quality can be solved; the real question, both from a technical and economic viewpoint is the extent to which they need to be solved.

The interconnection of customer-owned generating capacity represents a new phase in operations for the utility industry. A great deal of education and communication will have to take place between both the utilities and their customers before distributed generation capacity, particularly small wind machines, become commonplace.

7
OWNING A WIND MACHINE

Major zoning issues, wind machine safety and environmental impacts, educating the local zoning board, installation, operation and maintenance.

When Wayne Dingerson tried to install a wind generator in his property in the Denver, Colorado area, sixty of his neighbors signed a petition of protest and threatened a lawsuit to prevent the installation. Wayne gave up his plan saying, "I've got to live with these people. I'm sure I could have found a legal way around the county's height restrictions. I don't think the problem was the rules or regulations."

"I felt the opposition was unreasonable," he continued. The hurdle is not the government; the obstacle is the people. They have to be educated."

This story is common. You can spend a great deal of time researching wind energy, selecting a machine, and negotiating with the utility. Later, as you are about to become an independent energy producer, you encounter stiff neighborhood opposition. In the past, economics was the largest single obstacle to using small wind machines. Now, as the cost of utility electricity rises and more wind machines are being installed around the country, the economics have become attractive in many areas. But the fears and concerns of neighbors and the local zoning boards have become the formidable hurdle.

Many of the fears about small wind machines have been overstated. There appears to be two major reasons for this. First, the technology is new, at least in suburban and urban environments. Second, most literature on the environmental impact of wind machines relates to large utility-scale machines whose blades range in diameter from 100 to 300 feet. These documents are not applicable to the smaller machines. The environmental impact statements cover every possible concern, and early experiments with large machines produced problems such as television interference and noise.

The next part of this chapter discusses the major issues planners, zoning commissions, and consumers must assess to determine the risks and benefits of small wind machines. This discussion of major zoning issues, wind machine safety, and environmental issues is intended to allay unnecessary concerns and to address valid problems.

MAJOR ZONING ISSUES

Tower Height

As we have seen, siting plays a critical role in the performance of a wind machine. High towers are required to allow a clear access to the wind, so the machine can perform efficiently. Turbulence from wind passing over buildings, trees, and land features drastically decrease the performance of a wind machine installed close to the ground. Turbulence also increases stress on the wind machine, jeopardizing its safe performance and reducing its useful life.

Tower heights for small machines, usually range from 40 to 80 ft. In a few locations, 100 ft towers have been used because of nearby wind flow obstructions. The building height restriction of 35 ft in many residential areas could effectively prohibit the economical use of wind machines. As a Planning Advisory Service memo to the American Planning Association pointed out, strict application of such restrictions to wind systems, in effect, prohibits the use of wind energy.

A more sensible approach was proposed by a witness at a Board of Adjustment meeting in Boulder County, Colorado, who asked that the board include wind machines in that county's zoning resolution which states that:

> The height limitations of this resolution shall not apply to church spires, belfries, cupolas, penthouses, or domes not used for human occupancy, nor to chimneys, ventilators, skylights, water tanks, silos, parapet walls, cornices without windows, antennas, or necessary mechanical appurtenances usually carried above the roof level.

The Board voted in favor of his proposal and decided that wind machines are mechanical appurtenances and are therefore exempt from the building height restrictions up to 80 ft in height. Many other municipalities have modified their height limitations to allow the installation of small wind machines.

Setbacks

Setback requirements in local zoning regulations refer to the distance a structure must be from neighboring property lines. Setback requirements for a wind machine can be determined by a variety of factors:

- the consideration of aesthetics
- safety issues
- architectual review policies
- master plan requirements for future street widening
- allowance for wind access
- routine residential setback requirements for lot sizes

Communities may want to focus on one or several of these criteria for developing setback requirements for wind machines. In considering setbacks due to safety considerations (such as possible blade throw), communities may want to categorize wind machines into three groups with appropriate setbacks for each group, as recommended by Paul Wendelgass, a planner with the New York State Energy Office. In his discussion of the problem of blade throw, Wendelgass points out:

> A more useful approach to this issue (blade throw) would be to distinguish between machines which have a performance record which allows assessment of the likelihood of such a failure and those which do not. The most readily available means to make this distinction is to separate machines into categories of *ex-*

perimental, prototype, and production. Experimental machines are wind systems which are the first of their kind and their use constitutes an experimental testing of a new concept or design. Prototypes are machines in the next stage of development; construction of a limited number to test operations under field conditions. Finally, after this stage, wind machines enter the production phase, in which the manufacturer produces a significant number on a continuing basis.

Wendelgass suggests the most stringent setbacks be applied to experimental machines, less stringent for prototype models, and the least constraints on production wind systems. In the rare event of a blade throw, setbacks do not completely protect surrounding properties unless the distances required are 500 ft or more. Such distances would effectively prohibit wind machine installations in most areas and discriminate against wind machines in an environment where far more dangerous hazards exist.

Wind Rights and Wind Access

The matter of wind access or wind rights is related to tower height and setbacks. Who owns the right to the free-flowing wind both upwind and downwind of the wind machine? The wind machine owner is interested in maintaining an unobstructed wind flow for the safety and performance of his machine. Building and planting activity on neighboring properties could interfere with the wind flow. Resolving the question of who owns the wind involves land-use restrictions, property values, and the purchase of negative easements near a wind system. High concentration of wind machines are not likely to occur in urban settings. The problem of excessive turbulence present in the urban environment might exclude the practical siting of wind machines in many areas. Wind rights have not been addressed in proposed zoning regulations as they have only recently been identified as a concern. It is likely that in most cases, the question of wind rights will not be an issue.

Aesthetics

Aesthetic judgements are highly subjective. Personal taste, income, individual values, property values, and education are involved in the issue of the aesthetics of wind machines. If small wind machines are considered comparable to utility power transmission towers, street lights, telephone poles and similar objects, this may not be an issue. In places where these sights are common, public reaction to a small wind machine may be minimal. Prudent siting of a wind machine would diminish its visual impact. Nevertheless, members of the community should decide collectively about the aesthetic implications of wind machine installations. Local zoning regulations sometimes consider aesthetics and so reflect the community's values.

There are really two issues involved, and they can easily become confused. The basic issue is the mere presence of any man-made object of considerable height at the site selected. The other issue is the actual appearance of the system. Is it "too large" for the location? Is it "untidy" or "clumsy" looking? Does it look too modern in an area with older houses? The situation is the same as the problems faced by individuals who wish to build unusual houses in exclusive neighborhoods. There are no rules for resolving the issue of aesthetics, and its impact will be extremely site specific.

WIND MACHINE SAFETY AND ENVIRONMENTAL ISSUES

Blade Failure

Calculating the distance a wind machine blade would travel in the event of a failure is a complex task because there is no way to typify all wind machines or all environments in which they could be installed. There is no way to develop a generic blade travel analysis because there are no generic wind machines. Blade configurations, braking and feathering schemes, pitch controls, and other operational monitoring schemes differ from one machine to the next. Blade travel potential is also highly dependent on local turbulence, prevailing wind direction and strength, and other site-specific factors.

Commercially available, electricity-generating wind machines are different from the low rpm mechanical output water pumpers that dot the American landscape. Electrical output machines incorporate a wide variety of features to make them more efficient and safe. Computer-assisted structural analyses allow stress reductions which increase machine life and reliability. Brakes, blade pitch controls, and rotor axis controls are employed in various designs to protect the rotor and blades in high winds. Many machines are designed to withstand winds in excess of 100 mph.

Despite this advanced technology, the issue of blade or rotor failure continues to be raised in zoning hearings. Various generic blade travel analyses have been conducted, but each has been met with controversy. It is impossible to predict the aerodynamic forces on the blade, the manner of blade failure, and the blade orientation at the time of failure. A thrown blade may land at the base of the tower, or 500 ft away.

A properly engineered and designed blade on a well-designed wind machine with protective controls will have very low probability of failure. Also, blade failure is usually associated with mechanical problems or the presence of extremely high winds. Blade failure in low winds may be signaled by vibration or noise, in which case the machine could be "shut down" or secured. In high winds, hurricanes, tornados, or severe thunderstorms, it is unlikely that people will be moving about near a wind machine. The possibility of a blade throw is analogous to other things that break in high winds, such as tree limbs or entire trees, large signs, or utility poles.

Ice Throw

Questions about the potential of wind machine blades throwing ice or snow have been raised by a handful of communities in the northeastern United States. Because the blades are small and their surfaces smooth, ice and snow accumulation is not a problem. Any snow or ice that might build up on a wind machine blade would destroy the aerodynamics of the blade and prevent it from rotating very rapidly until the ice broke loose and fell to the ground.

Tower Failure

The failure of a wind machine's tower has historically been a rare occurrence. If local building codes specify that wind machine tower meet the requirements prescribed by the Uniform Building Code(UBC), there should be no unnecessary concern over the possibility of tower failure.

Not all communities have adopted the Uniform Building Code. Local building codes may require full or partial compliance with any one of the four major building codes (Uniform Building Code, National Building Code, Basic Building Code, or Southern Building Code) depending on the judgement of the building department. Applicable sections of the UBC are adequate for wind machine tower in that they stipulate calculations for structural phenomena associated with towers such as overturning moment, uplift, and load-carrying capacity. Most wind machines use standard towers. The standards ensure the structural integrity of towers. Wind system manufacturers and tower manufacturers can readily calculate the loads with the towers must withstand under the most extreme conditions. They have experience designing towers to endure these loads with an acceptable margin of safety. As a result, towers properly designed and built for a wind system are extremely unlikely to fail at their base.

Local ordinances governing wind machine towers have generally taken one of two approaches. Either the tower must be certified by a registered engineer, or it must be erected at a distance greater then its height from the property line. Because the wind machine and the tower interact in a dynamic fashion, many manufacturers specify the tower is appropriate for their machine, or provide the tower as part of the installation.

Tower Foundation

Tower foundation requirements also fall under the jurisdiction of the local building codes. This is particularly important in areas subject to flood, earthquake, or ground settling. Building code inspectors are aware of such local features, and local contractors can readily comply with specific requirements.

Tower Access

Almost every wind machine zoning ordinance or policy being drafted includes a provision to keep people from climbing the tower. The hazard of falling from the tower and the danger of contact with high-voltage equipment make this necessary. A variety of approaches have been taken, including installing anti-climb shrouds (smooth metal covers) over the bottom 12 ft of the tower, or removing climbing apparatus up to 12 ft above the ground on tubular towers. Fencing around the base of the tower has also been suggested in some areas. This may not be as effective as a permanent sign reading:

WARNING: ELECTRICAL SHOCK HAZARD

Electromagnetic Interference (EMI)

Wind machines can interfere with electromagnetic waves, television and FM radio, microwave communications, and aircraft navigational systems. Two possible causes have been identified: rotating blades which can scatter electromagnetic signals, and line interference caused by the wind machine's generator or inverter.

The potential for electromagnetic interference associated with wind machine blades is closely related to system and rotor solidity (ratio of blade surface area to total area swept by blades). A wind system with a small rotor diameter would have much less EMI potential than the larger systems (over 100 ft in diameter) for which EMI data are available. Small wind-machine-caused EMI has never been observed or reported, even though thousands of these machines are in operation. Many metal bladed wind machines are used around the world to power remote communications systems. These have been installed very close to signal reception and transmission points, with no adverse effects reported.

Although no incidents of interference associated with small diameter wind machines have been reported, many zoning ordinances provide for the modification or removal of any wind machine which causes interference. These provisions were enacted as a result of the EMI problems in Boone, North Carolina during the preliminary operation of a large experimental wind machine. The 2MW wind machine in Boone, called the Mod-1, has a rotor diameter of 200 ft. Television interference from the Mod-1 occurred in a roughly circular area around the machine where the blade-reflected interference signal was strong enough to be picked up on TVs. Data is not available for the complete range of small machine diameters, but experience shows that there is no interference beyond a few feet of the rotor.

Electrical Safety

The local electrical inspector has jurisdiction for electrical requirements for wind machines. Like many building codes, electrical codes are not the same in all states or communities. The local electrical inspector is responsible for outlining requirements for labeling, grounding, wiring, and other electrical safety considerations not under the jurisdiction of the utility to which the machine is connected. Requiring that wind machine installations comply with local electrical codes should be adequate for local authorities to be assured of safety, and not too difficult or costly for the wind machine owner.

Noise

Noise is frequently one of the first issues raised in discussions about wind machines, but the question is easily dismissed. Reporting on tests and analyses conducted at the Rocky Flats Test Center, A. C. Hansen made these observations on the subject of noise from small wind machines:

> Most cities, towns, and states have noise control laws or ordinances to protect communities from undue hazard or nuisance from noise sources. In addition, federal regulations apply to special noise sources such as aircraft. Most ordinances have a section dealing specifically with traffic noise and another section dealing with general noise sources. Wind machine noise falls into the general noise category in the vast majority of cases.
>
> The graph in Figure 7-1 shows a number of noise levels in decibels and gives an indication of familiar sounds. Note that traffic noise, one of the most common and annoying residential area noises, is over 80 dB and that typical office noise levels are approximately 50–60 dB. It can be concluded that wind machine noise need not pose a barrier to wind machine use.

Wind machines make noise only when there is wind. Because trees, buildings, and other objects also generate noise when there is wind, the background noise helps mask any noise from the wind machine.

Many noise ordinances specify that noise levels should be measured during calm periods. Because wind machines would not operate at such times, these ordinances are difficult to apply to wind machines. In fact, if noise levels at rated output (usually in winds over 20 mph) are written into ordinances, they can be discriminatory. For example, if 50 dB is arbitrarily chosen as wind machine limit, it would be quite

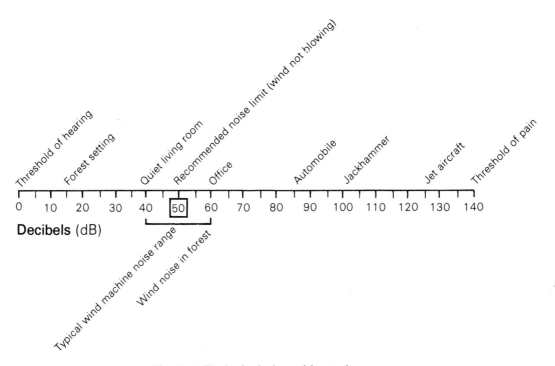

Fig. 7-1. Typical wind machine noise range.

noticeable on a calm evening. But if the wind is blowing through trees or shrubs at a speed of 20 mph, the ambient noise level would be much above the 50 dB level. For wind machine noise ordinances to be non-discriminatory, they must include consideration of background noise as well as wind machine noise.

Bird Safety

There is concern for the welfare of birds flying near wind machines. Tests have shown that the operation of even large wind machines pose no significant danger to birds. The Battelle-Columbus Laboratories conducted a two-year study on the biophysical impacts on the first Department of Energy sponsored large wind machine. During the study, birds flying near the rotor flew away from the blades. Others flew in a straight line between the rotating blades without incident. Only one dead bird was found near the 125 ft diameter wind machine in four migratory seasons.

These results are meaningful since this machine is much larger than the typical residential machine, and the test facility is located along a major flyway for migrating song birds. The conclusions of this and other studies are that during the day the birds can see and avoid the rotor. Birds usually fly at night only when migrating, at altitudes above 500 ft. It is commonly agreed that wind machines are of little threat to birds.

EDUCATING THE ZONING BOARD

Because the technology is relatively new, most local zoning laws make no mention of wind machines. Therefore, it is often unclear whether wind machines should be allowed by right, special permit or exception. Some communities have amended ordinances to specifically address the question of wind machine use.

Many more are in the process of creating such amendments. Others require special use permits which allow for the review of the application by the zoning commission. Still others are evaluating each wind machine installation on a case by case basis through the *use variance* process.

Because there are a number of avenues to obtain the necessary permits and licenses to erect a wind machine, including changing the ordinance, obtaining a *special use* permit, or simply obtaining a variance for an existing ordinance, resolving zoning issues before installing a wind machine can be a confusing and time-consuming process for the potential user. The following steps will help you plan a course of action and reduce the cost and effort required to obtain consent, approvals, permits, and variances that may be necessary to install a wind machine in a residential area.

Step 1: Learning the Issues

The most important first step is to learn everything possible about the machine and the issues the proposed installation can raise. A wind machine, unlike the automobile, is new to populated areas. Unlike a washing machine, it is not used in an out-of-the-way place. Because it is installed out of doors and must be in a prominent location, neighbors are likely to feel they will be living with a wind machine even if they don't own it. The potential wind machine owner should carefully study all potential issues.

Step 2: Soliciting Help from the Dealer

The dealer should supply much of the information necessary to answer questions like: How much energy will it produce? How tall will the tower be? What kind of safety record does it have? Is it noisy? The distributor can provide photographs, drawings, and case histories of similar machines installed elsewhere. Information on the safety of the machine, tower integrity, drawings and schematics will help people understand the operation of the machine. A good dealer will be your best ally because he or she has an economic interest in the decision of the zoning board, which may set a precedent for a large portion of the area the dealer serves.

Step 3: Polling your Neighbors

Once you are aware of the issues and have learned all the facts about a specific machine, poll your neighbors to get their feelings regarding alternative energy devices in general and wind machines specifically. Many neighborhoods have homeowners associations which may provide a forum for this discussion. Also, neighbors should be notified of any variance applications as local boards often base their decisions on the testimony of surrounding property owners. Neighbors appreciate learning of a proposed installation from the wind machine owner first. In addition you might want to take a few neighbors to a wind machine installation so they can hear and see a system for themselves. Showing slides or photos of the wind machine may allay concerns about aesthetics.

Step 4: Identifying the Zoning Barriers

The next step is to visit the local building inspector and zoning department to determine if a building permit is required and what zoning ordinances cover the proposed site. The building inspector will probably have concerns about the integrity of the tower and the safety of the system. Any information that can be furnished to answer these questions will be helpful. Most wind machines and towers have been tested under various stresses and loads, and the results may be available. If a warranty on the tower is available, this should be shown to the building inspector. At this time, determine if the system complies with the local electrical codes.

The zoning department may identify setback, height, or other zoning requirements. If any of these are too restrictive, you can appeal to an appeal board.

Step 5: Contacting State Energy Office or Local Energy Groups

These groups have information on alternative energy technologies, and are another source of local aid and information. They might have information about wind systems in other parts of the state as well as strategies for obtaining the proper permits.

Step 6: Preparing for a Public Hearing

Some background work must be done prior to public hearings before the zoning board, appeals board, or other pertinent groups. You should keep a record of the names and telephone numbers of individuals who supply information during this process, because they may be needed at the hearing. While zoning boards differ from one locale to another, these are typical steps to prepare for a hearing:

1. The best source of information about a zoning hearing is usually the zoning staff itself. Do not treat them as an adversary. Visit them early in the process to find out what will be required.
2. If you are appealing a negative decision before a board of appeals, find out what other boards reported to them and see if you can obtain a favorable recommendation from these other boards, all are usually asked to make recommendations to a board of appeals.
3. Search through the local records to find out how the zoning or appeals board has resolved other similar applications for height or use variations.
4. Contact your local utility and/or state utility commission and verify that your system complies with their requirements and that they will approve interconnection with the utility grid.
5. Contact the wind system distributor or manufacturer for test results, drawings, or photographs of the machine.
6. Demonstrate compliance with all applicable building and electrical codes. It may be necessary to have a professional registered engineer prepare a drawing showing:

    - tower foundation dimensions
    - foundation materials
    - method of attaching the tower to the foundation
    - wind system dimensions and characteristics (rotor diameter, tower heights, system weight, electrical schematic)

7. Contact the neighbors who approve of the installation to have them submit letters indicating that they have no objections. This step is not usually required but can be helpful.
8. If necessary, prepare a map to direct board members to the site in question.

It is important that any documents presented to the board be in the most professional form possible. Type your information and put it together in a folder.

Step 7: Making the Presentation

When presenting a variance case before the zoning board or board of appeals, stress why the variance should be granted. Board members are probably looking for reasons why they should grant a variance. Present your case as a routine application, not a "cause" or test case.

The board will ask questions about the height of the tower, safety of the machine, electrical characteristics, environmental impact, and other issues discussed in this chapter. The applicant should anticipate these questions and have answers preprared.

Step 8: Appealing a Negative Decision

If the application or variance is denied, there normally is a short filing period for appeals, ranging from 10 to 65 days. Another way to obtain the necessary permits is to change the local ordinance. If you can encourage the town to make wind systems a permitted use with exemptions from the normal height restrictions, a variance would not be required. A new ordinance could be drafted that would still provide a fair review of proposed installations.

Obtaining the necessary permits and approvals will become simpler as more wind machines are installed around the country, safety standards are developed, and towns adopt ordinances that include wind machines. Until then, local boards, acting on the existing information on wind machines, must fulfill their responsibilities to protect the public health and welfare. You must provide them with the information.

Because of the deregulation of oil and natural gas, the cost of wind energy will become more competitive with conventional energy sources. This will affect public concern about wind machines. The visual impact of wind machines will be compared with radio tower, TV antennas, street lights, or other tall objects on the skyline. Their presence in neighborhoods may become as common as the automobile, and might become a source of community pride. The issues and approach described here should help the wind system applicant and the zoning board. Local officials should weigh both the benefits and risks of wind machines, and judge the usefulness of this energy alternative for their community.

INSTALLATION

Installing a wind machine is a major construction project and individuals are urged to rely on the wind machine distributor to perform this task. Relying on professionals will ensure the installation is technically correct, and that the wind machine is safe and efficient (Figure 7-2 and 7-3).

The wind machine owner should discuss the installation procedures with the distributor and be present during erection. If questions arise, the installer and the owner can confer and avoid later difficulties.

OPERATION

Once installed, operating a wind machine is simple and straightforward. They require little maintenance, but like automobiles, preventative maintenance is less expensive than repairs. The distributor should pro-

Fig. 7-2. Hoisting a wind machine.

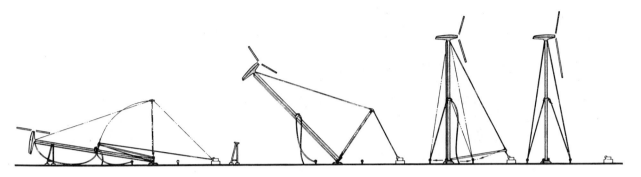

Fig. 7-3. Erecting a wind machine with a winch.

vide an owner's manual which describes the features of the machine, a parts list, necessary drawings and diagrams, and a maintenance schedule.

The distributor will teach you how to operate the machine, including how to shut off the machine for periods of maintenance or inspection. If unusually high winds are expected from a storm, weather front, tornado, or hurricane, the machine should be shut down. The few extra kilowatt-hours that would be obtained are not worth the risk of damaging this expensive piece of machinery. When in doubt, shut it down.

Any abnormal occurrence, unusual noises, dimming lights, oil or grease leaks—should be reported to the distributor at once. In most cases, you should shut the machine off until it is inspected.

To verify the original calculations of the economics of the machine, wind speed and energy production statistics must be collected. Two instruments are adequate to collect this information: an anemometer which can be mounted on the tower; and second, a kilowatt-hour meter. If the anemometer is mounted on the tower, remember it will always be obstructed by the tower from at least one wind direction. During the first year of operation, these instruments should be read frequently, perhaps monthly. If they are substantially different from what was predicted, the distributor should inspect the machine.

MAINTENANCE

An average automobile may be driven at 50 mph for 100,000 miles—2,000 hours of operation. The car might be serviced every 5,000 miles or 100 hours of operation. Wind machines are designed for 15-20 years of trouble-free operation. There are 8,760 hours in a year, and if the wind machine operates only 25% of the time, it will have as many hours of operation in one year as a car does after 100,000 miles.

Maintenance should be a major consideration when selecting a wind machine. Consider the following factors:

- Maintenance history of major components
- Frequency of scheduled maintenance
- The number and types of tools required for maintenance
- The availability and cost of replacement parts
- The availability of factory-trained maintenance personnel

The air mass flowing past a wind machine is full of dust and grit which will penetrate the machine and affect the bearings, and inspecting the other components is the best way to prevent rapid wear from such environmental conditions. Look for evidence of rust, loose wires, and worn parts during periodic inspections.

The blades are exposed to hail, rain, ice, dirt, and even rocks. Regularly inspect the blades and their attachments. Wooden blades may need fresh paint. Fiberglass or metal blades may need similar maintenance to their surface. Deterioration of the surface can reduce blade life and sharply reduce the machine's energy production.

Wind machine vibration causes bolts and nuts to loosen, parts to fatigue and fail, and wires to break or disconnect. Check each of these parts during the regular inspection program. Give your machine a thorough inspection once a year. In the early years of operation, inspect the machine after extreme weather.

This may sound like a lot of work, and it is. But almost no other machines run for 15 or 20 years, 365 days a year, 24 hours a day, in a completely open environment.

8
ONE FAMILY'S EXPERIENCE

"We did it for the kids. So we could have something for the kids. The kids are energy users now and in the future. We wanted them to grow up knowing that they cannot deplete an energy source without renewing it. Especially now, their generation . . . they've gotta learn that they must use things that are reusable. This society that throws everything away is gonna kill everybody."

So says Annie Roth, mother of three, as she gestures toward the wind machine in their back yard. Her husband, Carl, a jet engine mechanic, agrees: "These kids, our kids, aren't gonna have anything by the time they grow up. When we're gone, we wanted to leave them something."

Carl and Annie Roth became interested in wind energy for a variety of reasons. "We live right next to the ocean. I knew without even taking extensive wind measurements, that we have a lot of wind. The sea breezes. The land breezes. The night wind. The morning wind," Carl explained.

"We use about 800 kilowatt-hours of electricity a month." Annie interjects. "Two years ago we put in a new oil-burner for hot water, so we didn't consider solar water heating. We wanted a renewable energy source that would produce electricity."

Their home in Manomet, Massachusetts, near Plymouth and Cape Cod, is only minutes from the Atlantic Ocean, and ironically from Plymouth One, a nuclear reactor built by Commonwealth Edison to serve Massachusetts residents.

"We know all the dangers of living this close to that plant," Annie declares. "I grew up here and people know me and they know my parents. I may have a reputation for being outspoken, but I've known these people all my life." Annie, before the children became her full-time job, was an anesthesiologist's nurse.

A few years ago, Annie was reading a special issue of *National Geographic* on the energy crisis. The graphic, factual article described our planet's dwindling resources. She prompted Carl to read it, and together they began to explore alternative sources of energy. The location of their house helped them decide on wind energy.

"Before Christmas, we went to an alternate energy show at the mall. There was a model wind machine there on the ground where you could look at the parts." A local distributor displayed his wind machine and Carl spoke to him at length. The Roths collected all the manufacturers' brochures and got a list of distributors of wind machines from the American Wind Energy Association. In the months that followed, they sent away for all the information they could and they haunted the local library, looking for literature on wind energy.

"We wanted to find out all we could about the various machines. How they looked, who made them, how much they cost," Carl said.

Annie said, "and we went to the library here and to all the ones down the Cape. We just couldn't find much information on wind energy. So we made calls and wrote letters. We visited manufacturers and distributors in the area and asked them to explain their products."

The Roths immediately dismissed one machine because a person living near them owned it and told them he was unhappy with its performance. One Sunday afternoon, they went to visit one of the local manufacturers. "We saw the machine out there. It was love at first sight." Carl smiled. "The lights of the building were run off the machine. It was quiet. I decided right then and there, that was the machine we'd get."

The Roths had done a great deal of research on wind manufacturers and the products they offered. "We went back to the manufacturer on the following Monday. They were holding a class on wind energy when we walked in to find out how much the machine cost. The session was interesting and we learned how much we already know . . . our homework was paying off. A woman there asked about mounting the wind machine on her house, and she was disappointed to learn that she couldn't because of possible vibration problems."

"Most of the people there were very well-to-do and owned summer homes down on the Cape. One, a doctor from Connecticut, couldn't care less about learning about wind energy. His attitude was, 'Here's the money and here's my address. Put up the machine and send me a bill.' He wanted the company to take care of everything and was willing to pay the price."

After the class, the Roths made an appointment to have a company representative visit their house. They had already decided which machine they wanted and nothing could deter them. While they waited for the company to do a site analysis, Carl called the engineering department of Commonwealth Edison and asked if they had information on wind patterns around Plymouth One.

"I knew that in case of an accident, a meltdown or whatever, they'd have to know which way the wind would blow . . . they'd have wind information to measure pollution and dispersal of the fallout."

CommEd had very extensive wind records. And, to the Roth's surprise, they were happy to share the information. "They sent us the results of years of testing. We were right—we have an ideal wind site. We have a 14 mph annual average wind speed, and the windiest months are the months we use the most electricity."

The utility was cooperative in all phases of their project. "They gave us the extra meter out there, the one to measure how much power we feed back to them. We called their engineering department a number of times and explained what we were doing. They were very interested. Maybe curious is a better word. One of their engineers came out twice to make sure that this meter was installed correctly. The first time they sent an installer, he didn't know how to put it in. So, we just called CommEd again. The engineer came out. He promised us we'd have it installed in two days, and it was," Carl said.

"CommEd wanted to see how these wind machines would affect their operations. So, we've had a lot of help from them and no extra charges. But I suppose when these machines become commonplace, they'll have a regular procedure and a charge for the extra meter."

"We found that there was no such thing as 'useless' information. We even called the FAA (Federal Aviation Administration) to see if the tower would be considered a hazard to airplanes. They told us that unless you are within a mile from the end of a runway, you don't have to have lights on the tower. And, if it's less than 200 ft high, it does not have to be registered with them," Carl said.

Still, the Roths experienced a number of real setbacks and disappointments. They worked hard to get their machine installed.

"The zoning regulations would not allow a structure higher than 30 feet in our neighborhood, so we applied to the zoning board for a variance. Our first application was denied. We weren't surprised, we were ready for the refusal, from what we've read and heard from other people who installed wind machines."

"We put together booklets and gave them to the Appeals Board. These booklets had copies of everything we'd learned, all the specifications of the machine, all the safety features, test results, wind data, anything we could think of to include. The Board wasn't prepared to deal with this. They tried to stall us as long as they could," Carl said.

"Now, they have a Windmill Committee to deal with the problem," Annie added.

"When we started, the town didn't have any regulations for wind machines. The booklets were a tremendous help. The town kept them, and I think the Windmill Committee is using them as a model."

"We were surprised that the distributor didn't help us more. He just didn't take us seriously. Luckily, we knew what we wanted and had done enough research. We had the test results of that particular machine. The utility had already told us that we'd have no problem getting the machine hooked up to the grid, but they still hadn't offered us a price for our extra electricity. They said they wanted more information. We wrote our insurance company, and they assured us that our homeowner's policy would cover any liability incurred as a result of the windmill. We also contacted our neighbors about the project. We knew we would have problems with the Zoning Board and the appeals, but were ready for them."

The Roths got help from unexpected sources. A journalist was at the same building on the night of their first meeting with the planning board, heard their plans, and became very interested. He followed the case and wrote articles about them.

"We may never have gotten the variance without this reporter's articles. We were upstairs at a Planning Board meeting and they were discussing our wind machine downstairs at another meeting. The reporter called up the next night and said he was at the Zoning Board meeting. He thought we were there too. We didn't even know about the meeting. The city was holding meetings about the windmill that we didn't know about. Besides the Zoning Board, there was a Planning Board and a Design and Review Board. We needed help to get the variance and he did a lot to inform the public about our situation. And he would call us and say, 'They're having another meeting, you'd better get down here pronto!'

"We went to the Zoning Board and were automatically rejected. They didn't know how to deal with it. Then, we went to the Planning Board and explained the system to them: the height of the tower, kind of foundation, the electrical plans. We brought them a copy of our booklet.

"If I had it to do again, after the initial application had been refused, I'd go to the board and say, 'OK, it's been refused, now which board will review it and when will they meet?' I really think they didn't tell us about the meeting because they were making the whole thing up as they went along. But the articles and the media exposure helped educate people. We got interested in energy through the media and the media helped us get the machine erected."

Annie says, "One of the fellows I worked for in Plymouth, an anesthesiologist, read the article in the paper, called to say that he was behind the project and asked if he could help in any way at the hearing on our appeal. He came to the public meeting, stood up and introduced himself as a doctor, and went on to list all the benefits of wind energy. It's not polluting, it doesn't harm the environment. He went on and on. His testimony gave respectability to the entire project."

After months of appeals, the Roths were granted their permit. They hadn't invested any cash in their project, only time and energy. "We didn't want to go back to the Zoning Board after putting down a $10,000 deposit on the machine, and have our plans thrown out the window. The distributor had told us that every town in the state is different."

After the Board of Appeals granted a special permit, things went smoothly. The bank provided a second mortgage on the house and the new payments have been less than the old payments and utility bill. The utility came out to the house and installed the extra meters.

"We still get electric company trucks stopping in front of the house, the guys sitting in the trucks, looking at the wind machine."

CONCLUSIONS

"Looking back," Annie says, "I think we should have had more help from the distributor. I thought we'd get more help from them with the Zoning Board. We were lucky that we didn't have any problems with the utility."

Carl added, "If I were to do it again, I would go right to the distributor and ask him how much he was willing to help. I got frustrated at times. But we spent a lot of time researching the subject, and that really helped us with the town. We had all the information they needed—ahead of time.

"Anybody thinking about buying a wind machine should check out as many manufacturers as possible, and they'd better understand what a machine can and can't do. A lady up the block told her husband, 'Well, if we lived next door to that thing, we'd have free electricity too.' " People just don't understand. It took us more than a year, from the time we saw the model on the mall, till the time the machine was installed. I can't even guess how much time we have invested in the project."

"We were lucky in another way—we knew we had a good site. If we hadn't known that, we'd still be collecting wind data," Carl said.

"As more and more people buy and install wind machines, distributors will be more responsive, and towns and villages will have procedures for zoning permits," added Annie.

"Now, we use it and wait. We can't exactly call up a neighbor and ask, 'how's your machine going after ten years? What are the problems you've had?' We're pioneers. I guess it's kind of scary. This thing will pay for itself in a few years. But, who knows what can happen by then?" Carl says, smiling.

9
POSTSCRIPT

Wind energy has the potential of leading a major social and technological revolution in the 1980s. This book has focused on wind machines that generate under one-hundred kilowatts of electricity and the problems a wind machine owner will encounter in these early days of the wind industry. The size of the resource is many times larger than the electricity consumed in this country. With the right combination of competing energy costs, wind machine costs, and available wind, it can be economically employed nearly everywhere. Investment bankers have predicted the wind industry will reach $2 billion in sales by 1990, and thousands will be employed in the production, installation, and servicing of wind machines.

The residential market, ultimately using the largest number of wind machines, may be the slowest to develop. Employing wind energy in remote locations, that is, those not serviced by normal utility lines, is perhaps the most rapidly growing application of wind energy (Figure 9-1). Wind machines are presently generating electricity in the mountains of Morocco, on artic ice floes, Venezuelan jungles, the Canadian Rockies, offshore drilling platforms, and Australian deserts. Remote applications typically have extremely demanding environments. Temperatures can range from +140°F to −75°F; there can be hail the size of golf balls, and winds exceeding 165 mph. For remote applications when the cost of servicing the equipment quickly surpasses the original price of the equipment, the operational requirements are also severe: unattended operation for up to two years; 99% availability; 20-year life; and less than one hour of maintenance per year.

Even with these severe requirements, wind energy systems tend to be less expensive on a life-cycle cost basis, than other remote power options, either conventional or nonconventional. In a properly designed system, reliability levels surpass the best available alternatives. The chart in Table 9-1 compares the cost of energy from a range of options. While this may be more than you'd ever want to pay for electricity for your house, it can easily be seen why wind is rapidly proving to be the most cost-effective and reliable technology for remote, industrial power applications.

Another important area of rapid development is the large-scale wind machines. Large machines are really large. Preliminary design is under way on a machine 420 ft in diameter, with an electrical generating capacity of 7,200 kW. This field of development is different from either the residential market or the remote power market. The largest of these machines are the largest pieces of rotating equipment ever built by man. The structural fatigue experienced by the blades is in orders of magnitude greater than other high-fatigued structures like airplanes or bridges. These level of fatigue place large wind machines on the frontier of materials engineering. Even though the really large machines are experimental, the eventual users of the technology, the utilities, are cooperating closely with government and private researchers. When fully developed, the in-

Fig. 9-1. Desert microwave repeater station.

Fig. 9-2. Offshore wind farm.

tegration of these large machines into utility generation capacity is likely to be rapid. When large arrays of machines, small or large, are grouped in a wind farm, the output rivals that of conventional power plants.

Another important area of future development may well be the location of wind farms offshore. The wind energy potential world-wide just off the coast is in many cases much greater than the land-based resource (Figures 9-2 and 9-3).

Once technology for the large machines becomes sufficiently mature, it is likely that many countries will attempt to base wind machines off their coasts. Holland, Britain, and the United States have all conducted studies for offshore wind power systems that show they are technically feasible and potentially very economical.

Water-pumping wind machines are likely to be as important in other parts of the world as they were in the American West. There have been many wind energy experiments around the world in recent years, particularly in developing countries. In the most successful of these, in South America for example, wind energy has brought employment, higher agricultural production, and a higher standard of living.

Although most of the development work in this country has focused on the generation of electricity, the future is likely to find wind machines producing heat directly through friction. This process approaches 100% efficiency, and recent studies indicate it could be the most economical method of heating buildings, particularly in areas where winds are strongest in the winter.

94 YOUR WIND DRIVEN GENERATOR

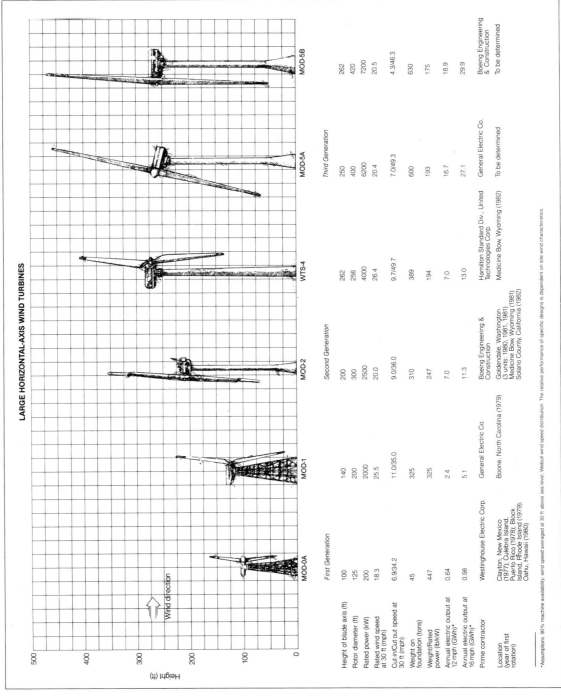

Fig. 9-3. Large horizontal-axis wind turbines.

Table 9-1. Ten-year cost comparison (400 watt load).

CAPITAL COSTS	WIND	THERMOELEC	GAS TURBINE	SOLAR	DIESEL
Generation Equipment	$9,800	$15,920	$18,854	$80,000	$ 5,560
Site Equipment Fuel Tanks, Installation, Tower, or Battery Housing	8,000	8,000	8,000	8,000	8,000
Battery Storage	6,000	1,500	1,500	6,000	1,500
INSTALLED COST	**$23,800**	**$25,420**	**$28,354**	**$94,000**	**$15,060**
OPERATING COSTS					
Annual Maintenance	$ 500	$ 750	$ 1,000	$ 500	$ 2,000
Fuel @ $1.50/gal.	—	6,570	4,098	—	3,022
Ten Year AO&M	5,000	73,200	50,980	5,000	50,220
Ten Year Life Cycle Cost	29,300	98,620	79,334	99,500	65,280
Ten Year kwh Production	35,040	35,040	35,040	35,040	35,040
COST PER KWH	**$ 0.84**	**$ 2.81**	**$ 2.26**	**$ 2.84**	**$ 1.86**

Remote power systems, large wind machines, wind farms, offshore arrays, water pumpers, and wind-driven heaters all offer tremendous benefits for mankind. Whether or not that potential is realized will depend to no small degree on the decisions made and actions take over the next few years.

SELECTED REFERENCES

These are references that were used in the preparation of this book. Many were used more than once, but they are credited in the chapter from which material from them first appears. The author is deeply indebted to authors of these publications, and the effort and knowledge they represent.

Chapter 1. Energy. The Basics

Annual Report to Congress, Energy Information Administration, U.S. Department of Energy, 1979
Small Wind Systems Application Analysis, JBF Scientific Corporation, U.S. Department of Energy, 1981.
Wind Machines 2nd edition, Frank Eldridge, Van Nostrand Reinhold Co., 1980.
Wind Energy Mission Analysis, Space Division, General Electric Co., Energy Research and Development Administration, 1977.
Rural Electric Cooperative Experience and Concerns with Interconnection of Small Wind Energy Systems, Lowell Endahl, (from) "Proceedings Small Wind Turbine Systems 1981," U.S. Department of Energy, 1981

Chapter 2. The Wind as a Source of Energy

Patterns in Nature, Peter Stevens, Atlantic-Little, Brown, Co., 1974
A Siting Handbook for Small Wind Energy Conversion Systems, Wegley, Orgill and Drake, Pacific Northwest Laboratory, 1978
Planning Manual for Utility Application of WECS, Park, Krauss, Lawler, and Asmussen, Michigan State University, U.S. Department of Energy, 1979.
Wind Energy Resource Atlas, Battelle Pacific Northwest Laboratory, U.S. Department of Energy, 1980–1981 (13 volumes).
Meteorological Aspects of Siting Large Wind Turbines, Hiester and Pennel, Pacific Northwest Laboratory, U.S. Department of Energy, 1981.
Vegetation as an Indicator of High Wind Velocity, Hewson, Wade, Baker, Heald, Oregon State University, U.S. Department of Energy, 1978.
Wind Energy Posters (series of three), American Wind Energy Association, Solar Energy Research Institute, 1981.

Chapter 3. Is Wind Energy Practical

Is the Wind a Practical Source of Energy for You?, DOE Rocky Flats Wind Systems Program, U.S. Department of Energy, 1980.
The Wind Power Book, Jack Park, Cheshire Books, 1981.
Wind Energy: An Introduction, The American Wind Energy Association, 1980.

Chapter 4. Choosing the Right Machine

Performance Rating Document (draft standard), The American Wind Energy Association, 1982.
First Semiannual Report: Volume II: Experimental Data Collected from Small Wind Energy Conversion Systems, DOE Rocky Flats Wind Systems Program, U.S. Department of Energy, 1978.

Chapter 5. A Wind Machine Investment

SWECS Cost of Energy Based on Life Cycle Costing, W. R. Briggs, DOE Rocky Flats Wind Systems Program, U.S. Department of Energy, 1980.
Managerial Finance: Principles and Practice, Steven Bolten, Houghton Mifflin Co., 1976.
Near-Term High Potential Counties for SWECS, Arthur D. Little Inc., Solar Energy Research Institute, 1981.
Small Wind Systems Technology Assessment: State of the Art and Near Term Goals, DOE Rocky Flats Wind Systems Program, U.S. Department of Energy, 1980.

Chapter 6. The Individual Wind System and the Utility

A New Prosperity, Solar Energy Research Institute, Brick House Publishing Co., 1981.
Rulemakings on Cogeneration and Small Power Production, Federal Energy Regulatory Commission, 1980.
Cogeneration and Small Power Production, Federal Energy Regulatory Commission, 1980.
Electric Utility Acceptance Criteria, Stiller and Shankle, Westinghouse Electric Co., prepared for North Wind Power Co., 1981, (unpublished).
A Handbook on the Sale of Excess Electricity by Industrial and Individual Power Producers under the Public Utility Regulatory Policies Act, Berger, Royce, Farley, Solar Energy Research Institute, 1980.
Study of Dispersed Small Wind Systems Interconnected with a Utility Distribution System, Curtice, Patton, Bohn, Sechan, Systems Control, DOE Rocky Flats Wind Systems Program, 1980.
Utility Concerns about Interconnected Small Wind Energy Conversion Systems, Bawn, Guerrero, Rocky Flats Wind Systems Program, 1980.
Issues and Examples of Developing Utility Interconnection Guidelines for Small Power Production, Lawless-Butterfield, Guerrero, Pykkonen, States, Rocky Flats Wind Systems Program, 1981.
Study of the Problems Associated with Interconnection of Wind Generators with Rural Electric Distribution Lines, O. Zastrow, for National Rural Electric Cooperative Association, (undated).

Chapter 7. Owning a Wind Machine

Wind Power for Farms, Homes, and Small Industry, Jack Park and Dick Schwind, U.S. Department of Energy, 1981.
The Technologies of Small Wind Energy Conversion Systems, R. L. Moment, Rocky Flats Wind Systems Program, U.S. Department of Energy, 1981.
Solar Law Reporter, Solar Energy Research Institute, quarterly periodical.
Small Wind Energy Conversion Systems: Zoning Issues and Approaches, Rocky Flats Wind Systems Program, U.S. Department of Energy, 1981.

Chapter 9. Postscript

Power for Remote Telecom Sites may be Blowin' in the Wind, Donald Mayer, "Telephony," December 21, 1981.
Wind Power: A Turning Point, Christopher Flavin, Worldwatch Paper 45, 1981.

GLOSSARY

This Glossary was prepared by Hugh Currin as chairman of the American Wind Energy Association's Terminology Subcommittee of the AWEA Standards Development Program.

Airfoil. A surface designed to create lift as air flows over it.

Anemometer. A device to measure wind speed.

Angle of attack. Angle between the airfoil chord line and the wind velocity.

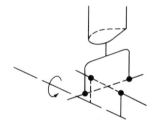

Articulated blade. A blade which is connected to its hub or central shaft through one or more hinge lines about which the blade is free to move. Articulation generally refers to the combined freedom about the inplane and out of plane axis. (See diagram.)

Aspect ratio. One-half the rotor diameter squared divided by the projected area of one blade.

Atmospheric boundary layer. A region of air close to the ground within which the major turbulence effects of surface roughness and heating are concentrated.

Availability factor. The ratio of time a WECS is available to produce power to the total time interval under consideration. (This is independent of wind speed.)

Axial thrust. Aerodynamic force exerted on a WECS rotor parallel to the wind velocity.

Battery storage. Storage of energy in the form of potential chemical reactions with this energy being recovered in the form of electricity.

Blade. Element of a WECS rotor which forms an aerodynamic surface to extract energy from the wind.

Blade deflection. Relative displacement of points in a blade due to blade loads.

Blade load distribution. A description of how blade loads are distributed along the quarter chord line.

Blade loads. Forces applied to a WECS blade.

Blade pitch. Angular setting of a blade measured as the angle between the chord line and a line tangent to the arc swept out, due to rotor angular velocity, by a point fixed at the radial station of interest. This angle is measured positive in the direction which will give a positive angle of attack.

Blade pitch control. A mechanism which uses blade pitch for rotor control (usually for overspeed and shut down).

Blade planform. A geometric description of a blade. (Usually includes twist, chord, and airfoil changes with changes in position along the quarter chord line.)

Campbell plot. A graph of a WECS system natural frequencies versus rotor angular velocity.

Capacity credit. The percentage of conventional generating capacity which may be eliminated from new requirements if a wind generator is added.

Capacity factor. Energy output of a WTG system divided by the energy output which would have been obtained if the machine had operated at maximum power output over the same time interval.

Chord. Distance along the chord line between the leading and trailing edges of an airfoil.

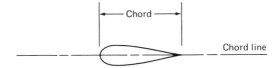

Chord line. Straight line constructed through an airfoil leading and trailing edges. (See above diagram.)

Collective pitch. The simultaneous change in pitch of all rotor blades by the same amount in the same direction.

Concentrator. A device of structure which collects energy in the wind and forces it, still in the form of wind energy, into a smaller area. (Usually refers to man-made structures but may also apply to natural formations.)

Configuration. Description of the basic elements and relative orientation of these elements in a particular, or generic, type of WECS.

Coning. Angle between the rotor plane and the blade quarter chord line. This term has meaning only when the quarter chord line is assumed straight. (See diagram.)

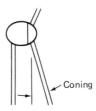

Constant speed load. A load which constrains its rotational motion to a constant value for all input torques of interest (i.e., synchronous generator).

Control system. WECS subsystem that senses the condition of the WECS or environmental parameters and depending on these conditions adjusts WECS operation to protect it or optimize output.

Cross wind. Component of wind velocity which is perpendicular to the mean wind velocity. Usually horizontal but can also be vertical.

Cup anemometer. A rotating drag device for measuring wind speed characterized by cups mounted on radial arms about a vertical axis. (See diagram.)

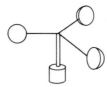

Cut-in wind speed. Wind speed at which a WECS system begins to deliver power.

Cyclic pitch. A consecutive change in pitch of each rotor blade from a minimum to a maximum, one cycle, each revolution of the rotor.

Dispersed systems. Term used to describe applications in which WECS are installed at many separate locations.

Down wind. HAWT configuration in which the rotor is down wind of the tower. (Wind passes first the tower, then the rotor plane.)

Drag. Aerodynamic force on a surface which is parallel to the airflow that is producing that force.

Drag device. Device which extracts energy from the wind by using primarily drag forces.

Drive train. WECS subsystem which transmits torque between the rotor main shaft and the generator or other load.

Electrodynamic wind driven generator. [I never saw one, have you?]

Fan tail. A secondary rotor used on some HAWTs to extract energy from a cross wind to change the yaw angle.

Feather. To pitch each blade of a rotor to a zero lift condition (normally used as a method of shut down).

Fixed coning. Describes a rotor which has no articulation in coning.

Flapping. Blade motion in the out of plane direction. This motion may be centered at a flap, articulation, hinge or distributed along the blade, distributed flexibility.

Flutter. The dynamic instability of an elastic body in an airstream.

Footing. Structure buried in the ground to support a tower.

Free coning. A rotor which is articulated to allow blade motion in the out of plane direction.

Free standing tower. A tower that does not use guys.

Fuel saver mode. WECS operating mode wherein energy produced is substituted for other forms of energy, thus saving fuel.

Generator. Machine or subsystem that transforms mechanical energy into electrical energy.

Governor. Control system, or part of, which controls rotor speed and/or torque to prescribed values.

Grid. Network to transmit and distribute electrical energy from point of production to point of end use. (Usually AC being tightly controlled to 50 or 60 cycles/sec.)

Gust. A positive or negative short term variation from the mean wind speed.

Gust factor. Ratio of wind speed averaged over a short period to the mean wind speed.

Guy. An element of a tower that is anchored to the ground and is designed to carry loads only in tension.

Guy anchor. Tower footing designed for guy connection.

Guyed tower. Tower designed to use guys for support.

Heat storage. Storage of energy in the form of a temperature differential.

Height diameter ratio. Ratio of height from the ground of the area centroid of the rotor swept area to the rotor diameter.

High speed rotor. A rotor whose tip speed ratio is five or more.

Horizontal axis wind turbine. A WECS whose axis of rotation is parallel with the ground and the mean wind velocity.

In plane. Denotes blade motion in the direction of the rotor plane.

Induction generator. A type of electric generator which runs above synchronous speed and draws all of its excitation (reactive power) from the system it is feeding into. (It generates at a constant power factor.)

Interconnected system. A WTG which is electrically connected to the utility grid in a manner that allows it to feed power into that grid.

Inverter. A machine which will convert DC electrical energy into an AC form.

Kite anemometer. A kite which has been calibrated to give quantitative wind speed data.

Leading edge. 1) The point on an airfoil which is the farthest from the trailing edge. 2) Area of blade surface which is the first to come in contact with the air flow.

Lift. Aerodynamic force which is perpendicular to the air flow that is producing that force.

Lift device. WECS which depends primarily on lift to extract energy from the wind.

Mean power output. The average power output of a WECS at a given mean wind speed based on the Rayleigh frequency distribution. Defined as:

$$\text{MPO}(\overline{V}) = \int_0^\infty P(V) \, F(\overline{V}, V) dV$$

where:

$\text{MPO}(\overline{V})$ = mean power output as a function of $\overline{V}$.
$P(V)$ = WECS power output as a function of V.
$F(\overline{V}, V)$ = Rayleigh frequency distribution as a function of $\overline{V}$ and V.
$\overline{V}$ = mean wind speed.
V = instantaneous wind speed.

Mean wind speed. Statistical mean of instantaneous wind speed taken over the time of interest, here being several minutes to a year.

Nacelle. A permanent covering placed over the generator, drive train, and other elements on top of a WECS tower to protect these from environmental conditions.

Natural frequency. The frequency at which a system will tend to vibrate without the application of a periodic forcing function.

Offshore WECS. WECS which are mounted on offshore platforms, flotillas, or buoys.

Out of plane. Denotes blade motion perpendicular to the rotor plane.

Overspeed control. The action of a control system, or part of, which prevents overspeeding of the rotor.

Pattern factor. An index which describes the status of units over a wide area, for example, a utility load pattern.

Peak wind speed. Maximum instantaneous wind speed occurring over the time under consideration.

Power conditioning. To change or modify the characteristics of electrical power. (i.e., DC to AC, or 12V to 32V).

Power curve. A plot of WECS power output versus mean wind speed.

Power density. Amount of power per unit area.

Power output. Useful power delivered from a WECS system at a stated mean wind speed.

Projected area. As seen from the direction of wind velocity, the area covered at any instant by the rotor blades. (Area solidly covered by the blades as opposed to the swept area).

Propeller anemometer. A HAWT whose output is used exclusively to determine wind speed.

Pumped hydro storage. Storage of energy potential in the form of a water head.

Quarter chord. The point on an airfoil chord line one-quarter chord back from the leading edge.

Quarter chord line. A line passing through the locus of quarter chord points of a rotor blade.

Quarter chord moment. Aerodynamic moment that causes a torque about the quarter chord point of an airfoil in a flow.

Radial station. An airfoil section of a blade a certain distance along the quarter chord line.

Rated power. The power output obtained from a WECS operating at its rated wind speed.

Rated wind speed. Wind speed at which the rated power is specified. (Usually the lowest wind speed which give maximum WECS output.)

Rayleigh frequency distribution. A mathematical idealization giving a ratio of time the wind blows within a given wind speed band to the total time under consideration. This distribution is dependent only on mean wind speed. Defined as:

$$F(\overline{V},V)dV = \frac{\pi}{2} \frac{V}{\overline{V}^2} \exp\left[-\frac{\pi}{4}\left[\frac{V}{\overline{V}}\right]^2\right] dV$$

where:

$F(\overline{V},V)$ = Rayleigh frequency distribution as a function of $\overline{V}$ and V.
$\overline{V}$ = mean wind speed
V = instantaneous wind speed
dV = differential wind speed band under consideration

Resonance: Potentially destructive condition where the frequency of the applied load equals the natural vibration frequency.

Rotor. A system of rotating aerodynamic elements attached to a single shaft that converts the kinetic energy in the wind into mechanical shaft energy.

Rotor axis. Axis of the rotor shaft through which power is extracted from a WECS rotor.

Rotor diameter. Twice the distance from the rotor axis to the outermost point on the blade.

Rotor plane. A plane perpendicular to the rotor axis passing through a point where the blade quarter chord lines intersect.

Rotor power coefficient. Power delivered to the rotor main shaft, rotor axis, divided by the power available in the flow field at a specified wind speed.

Rotor tilt. In a HAWT, the angle between horizontal and the rotor axis.

Rotor torque. The moment (torque) produced by a rotor about its axis.

Run of the wind. The length measurement of the wind that traverses a point in a given time interval.

Shut down. A safe condition for a WECS which is not its power producing mode but either 1) protects the WECS from unusual circumstances, or 2) protects those working on the machine.

Shut down wind speed. Wind speed at which the control system will shut down the WECS.

Site survey. The evaluation of a proposed or existing WECS site with respect to those parameters important to its operation.

Slip rings. Assembly used to transfer electrical power or signals from a rotating shaft.

Small wind energy conversion system. WECS system that outputs a maximum useful power output less than an equivalent of 100 kW. This power, stated here in kW, may be in any useful form (i.e., hp, Btu, etc.).

Solidity. Rotor project area divided by that rotor's swept area.

Stall. Aerodynamic condition where the flow over an airfoil is separated. (Loss of lift and increased drag are indications of stall.)

Stand alone inverter. Inverter that operates without being connected to a utility or other grid for frequency and/or voltage signals.

Storage: The saving or storage of energy produced at one time for use at a later time.

Survival wind speed. The maximum wind speed a WECS in automatic, unattended operation (not necessarily generating) can sustain without damage to structural components or loss of the ability to function normally.

Swept area. Area as seen from the direction of the wind that a rotor will, pass over during one complete rotation.

Synchronous generator. A type of electric generator which runs at synchronous (constant) speed and draws its excitation from an external or independent power source from the system it is feeding into. (It generates at variable power factor)

Synchronous inverter. Inverter which depends on a grid for timing and/or voltage level signals.

System power coefficient. Power output of a WECS system divided by the power available in its flow field at a specified wind speed.

System weight. Total weight of a WECS system including everything above ground level. (This includes tower, machine, wiring, etc., but excludes footings, guy anchors, etc.)

Tailvane. Vertical surface mounted on a WECS for control purposes. 1) some HAWTs use this for yaw orientation and control; 2) some VAWTs use this for coordinate cyclic blade pitch.

Tip speed. Linear speed of the point on a rotor farthest from the rotor axis.

Tip speed ratio. Tip speed divided by the corresponding wind speed.

Tower. Subsystem of a WECS that holds the rotor, or other collection device, above the ground.

Tower shadow. Aerodynamic wake created by airflow around a tower.

Trailing edge. 1) The point on an airfoil farthest down stream when operating at zero angle of attack, 2) area of a blade that is last to contact an element of airflow.

Turbulence. Rapid wind fluctuations.

Turbulence intensity. Ratio of the first standard deviation of wind speed to the mean wind speed.

Turn out of the wind. To yaw a HAWT so that its rotor axis is perpendicular to the wind direction.

Twist. Variation in blade pitch with respect to changes in radial station.

Upwind. HAWT configuration where the wind intercepts the rotor plane prior to the tower.

Vertical axis wind turbine. A WECS whose rotor axis is perpendicular to a horizontal plane.

WECS array. Three or more WECS at one location.

WECS site. Location that is being considered for installation of a WECS or has a WECS installed.

WECS system. Description of a WECS that includes all equipment for its operation. (This includes tower, footings, wiring, power conditioning equipment, etc.)

Wind energy conversion system. A machine that converts kinetic energy in the wind into a usable form.

Wind rose. A graph showing the average wind speed from each of the standard compass points and the percent time the wind blows from each direction. (Usually monthly or yearly wind averages and 16 equally spaced directions are used.)

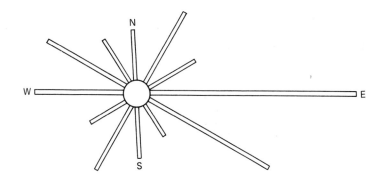

Wind shear. Difference in wind speed and/or direction with respect to differences in height.

Wind speed. Wind speed in a flow undisturbed by obstacles.

Wind speed duration curve. A graph which shows the distribution of wind speeds as a function of the cumulative hours that the wind speed exceeds a given wind speed in a year.

Wind speed profile. A graph depicting wind speed as a function of height above ground.

Wind turbine generator. A WECS whose output is electrical power.

Wind velocity. Wind speed and direction in an undisturbed flow. (A vector quantity.)

Yaw. Rotation of a HAWT about its yaw axis.

Yaw angle. Angle between the mean wind direction and the rotor axis.

Yaw axis. A vertical axis about which a HAWT changes its directional orientation.

Yaw control. A control system function that effects yaw angle or yaw rate.

Yaw orientation. A change made in yaw angle due to a control system action.

Yaw rate. Yawing angular rate.

Equivalent Terms

cut out wind speed = shut down wind speed
furling speed = shut down wind speed
HAWT = horizontal axis wind turbine
panemone = VAWT
SWECS = small wind energy conversion system
VAWT = vertical axis wind turbine
WECS = wind energy conversion system
windmill = wind energy conversion system
WTG = wind turbine generator

APPENDICES

AVAILABLE STATE INCENTIVES

This list was prepared by the firm of Arthur D. Little for the Solar Energy Research Institute.

ALABAMA INCENTIVES FOR SMALL WIND ENERGY CONVERSION SYSTEMS (SWECS)

INCENTIVE	RECIPIENT CATEGORY	PROVISIONS	RESTRICTIONS AND/OR PERFORMANCE CRITERIA	APPLICATION PROCEDURE	CARRYOVER PROVISIONS	DATES OF PROGRAM
Income Tax Incentive (none)						
Sales Tax Incentive (none)			NO INCENTIVES AS OF MARCH 1982			
Property Tax Incentive (none)						
Loans (none)						
Grants (none)						
Anemometer Loan (none)						

NOTES
Wind Energy Audits Not Required Under Federal RCS Program

CONTACT

Fred Braswell
Alabama Energy Management Board
3734 Atlanta Highway
Montgomery, AL 36130
(205) 832-5010

ALASKA INCENTIVES FOR SMALL WIND ENERGY CONVERSION SYSTEMS (SWECS)

INCENTIVE	RECIPIENT CATEGORY	PROVISIONS	RESTRICTIONS AND/OR PERFORMANCE CRITERIA	APPLICATION PROCEDURE	CARRYOVER PROVISIONS	DATES OF PROGRAM
Income Tax Incentive (credit)	Residential	Credit of 10% of installed system cost, up to $200. In addition to Federal credit.	None	Tax form notes additional form.	Program ends 1980.	1977-1980
Income Tax Incentive (credit)	Commercial Farms, Businesses	Credit of 35% of installed systems cost, up to $5000. In addition to Federal credit.	System must provide at least 10% of average thermal, electrical, or mechanical energy needs. Subject to energy or tax office audit.	Tax form notes additional form. SWECS owner provides supporting data.	Program ends 1980.	1977-1980
Sales Tax Incentive (none)						
Property Tax Incentive (none)						
LOANS (Alternative Technology and Energy Revolving Loan Fund)	Alaska residents	1978-1980 program: maximum $10,000 loan at 9.5% interest with 20 year payback. 1980+ program: maximum $10,000 loan at 5% interest. Loans deducted from income for Federal 40% tax credit.	Acceptability of proposal determined by revenue office.	4 page form plus supporting documentation required.		1978-1980 1980 – until funding used up.
LOANS (Renewable Resource Corporation Loans)	Any large Alaskan enterprise	Maximum of $550,000 available at 3% interest.	Applicants must have Alaska residency for at least 3 years.	Detailed proposal required.		1978 – no expiration date.
Grants (none)						
Anemometer Loans (none)						

NOTES

Income tax has been abolished for 1981 year. Legislature considering a rebate for all renewable energy systems purchased. Alternative Technology and Energy Revolving Loan Fund gave approximately two wind loans for $10,000 at 9.5% interest 1978-1980. Publicity was poor. New loan program has just been created.

Wind Energy Audits Authorized Under Federal RCS Program

CONTACT

Bob Shipley
Alaska Division of Energy and Power Development
338 Denali Street
Anchorage, AK 99501
(907) 276-0508

ARIZONA INCENTIVES FOR SMALL WIND ENERGY CONVERSION SYSTEMS (SWECS)

INCENTIVE	RECIPIENT CATEGORY	PROVISIONS	RESTRICTIONS AND/OR PERFORMANCE CRITERIA	APPLICATION PROCEDURE	CARRYOVER PROVISIONS	DATES OF PROGRAM
Income Tax Incentive (credit)	Residential	Credit of 35% of installed system cost, up to $1,000; reduced by 5% yearly starting 1984. In addition to federal credit.	Home built system costs do not include installation costs	Tax form notes additional form.	5 years	1978-1989
Income Tax Incentive (credit)	Commercial Farms, Businesses	Credit of 35% of installed system cost. No dollar maximum. Reduced by 5% yearly starting 1984. In addition to federal credit.	Same as above	Tax form notes additional form.	5 years	1978-1989
Sales Tax Incentive (exemption)	All SWECS owners	State sales tax of 4% not paid.	None	Not taxed at purchase.		1977-1981
Property Tax Incentive (exemption)	All SWECS owners	SWECS exempted from assessment for property tax until 1989.	None	Owner monitors assessment.		1978-1989
Loans (none)						
Grants (none)						
Anemometer Loan (none)						

NOTES

Director of Solar office knew of no credits/exemptions granted for wind.

Wind Energy Audits Authorized In Some Counties Under Federal RCS Program

CONTACT

James Warnock, Director
Solar Energy Commission
Capital Tour, Room 502
1700 W. Washington
Phoenix, AZ 85007
(602) 255-3682

ARKANSAS INCENTIVES FOR SMALL WIND ENERGY CONVERSION SYSTEMS (SWECS)

INCENTIVE	RECIPIENT CATEGORY	PROVISIONS	RESTRICTIONS AND/OR PERFORMANCE CRITERIA	APPLICATION PROCEDURE	CARRYOVER PROVISIONS	DATES OF PROGRAM
Income Tax Incentive (deduction)	All SWECS Owners	Deduction of 100% of installed system cost from taxable income. In addition to federal credit.	System must be expected to operate at least 5 years. Expenses must be incurred before 12/31/84.	Tax form notes additional form.	Until used up.	1979-1984
Sales Tax Incentive (none)						
Property Tax Incentive (none)						
Loans (none)						
Grants	Arkansas Residents	For "innovative energy systems."	System must come from well-known manufacturer or meet energy office approval. Applicant must not employ over 50 people.	Submit application and plans.		1979-1981 (extendable)
Anemometer Loan (none)						

NOTES

State Energy Office estimates 6 WECS in state
State Energy Office hopes to inaugurate anemometer loan program next year.

Wind Energy Audits Authorized In Some Counties Under Federal RCS Program

CONTACT

Nick Brown, Solar & Wind Project Coordinator
Arkansas Department of Energy
300 Kavanaugh Blvd.
Little Rock, AR 72205
(501) 371-1370

CALIFORNIA INCENTIVES FOR SMALL WIND ENERGY CONVERSION SYSTEMS (SWECS)

INCENTIVE	RECIPIENT CATEGORY	PROVISIONS	RESTRICTIONS AND/OR PERFORMANCE CRITERIA	APPLICATION PROCEDURE	CARRYOVER PROVISIONS	DATES OF PROGRAM
Income Tax Incentive (credit)	Residential (includes landlords not living on premises).	Credit of 55% of installed system cost minus Federal credit claimed (now 40%) = 15%, not to exceed $3,000.	SWECS Must: – produce at least 37.3 watts of electricity in 25 mph wind – be self-starting or have automatic start – have overspeed control or manual activator Manufacturer must provide 3 year warranty; installer must provide 1 year warranty.	Tax form notes additional form; submit with personal statement and/or manufacturer's plan.	Until used up.	1978-1983
Income Tax Incentive (credit)	Commercial Farms, Businesses	Credit of 25% of installed system cost of systems over $12,000, minus Federal investment tax credit claimed (now 10%) = 15%, not to exceed $3,000. Systems under $12,000 receive residential credit.	Same as above, plus: system's energy produced must be integral to commercial process.	Same as above.	Until used up.	1978-1983
Sales Tax Incentive (none)						
Property Tax Incentive (none)						
Loans (none)						
Grants (none)						
Anemometer Loans (none)						

NOTES: California legislature set a goal of 10% of state electricity to come from wind by year 2000; is encouraging installation of large WECS. State Energy Office estimates 12 SWECS put up by individuals.
Income tax credit is to stay at 55% including Federal.
If Federal changes, state will change.
Bill waiting for governor's signature which would raise credit up to 50% of system cost for commercial farms and businesses, including Federal credit.

Wind Energy Audits Authorized In Some Counties Under Federal RCS Program

CONTACT

Ron Kukula
California Energy Commission
1111 Howe Avenue
Mailstop #69
Sacramento, CA 92825
(916) 920-6023

COLORADO INCENTIVES FOR SMALL WIND ENERGY CONVERSION SYSTEMS (SWECS)

INCENTIVE	RECIPIENT CATEGORY	PROVISIONS	RESTRICTIONS AND/OR PERFORMANCE CRITERIA	APPLICATION PROCEDURE	CARRYOVER PROVISIONS	DATES OF PROGRAM
Income Tax Incentive (credit)	Residential	Credit of 30% of first $10,000 of installed system cost, not to exceed $3,000. In addition to Federal credit.	None yet.	Must obtain special tax form.	5 years	1980-1986
Income Tax Incentive (credit)	Commercial farms, business	Credit of 30% of installed system cost; no dollar maximum. In addition to Federal credit.	System must be installed between 1981-1986.	Must obtain special tax form.	5 years	1981-1986
Property Tax Incentive (exemption)	All SWECS owners	Exempts SWECS from assessment for property tax until 1990.	None	Owner monitors assessment		1979-1990
Sales Tax Incentive (none)						
Loans (none)						
Grants (none)						
Anemometer Loan (none)						

NOTES

Program is new. State Energy Office planning major publicity this fall.

Wind Energy Audits Authorized In Some Counties Under Federal RCS Program

CONTACT

Mark McRae, State Solar Officer
Colorado Office of Energy Conservation
1600 Downing Street
Denver, CO 80218
(303) 839-2186

CONNECTICUT INCENTIVES FOR SMALL WIND ENERGY CONVERSION SYSTEMS (SWECS)

INCENTIVE	RECIPIENT CATEGORY	PROVISIONS	RESTRICTIONS AND/OR PERFORMANCE CRITERIA	APPLICATION PROCEDURE	CARRYOVER PROVISIONS	DATES OF PROGRAM
Income Tax Incentive (none — no income tax in this state)						
Sales Tax Incentive (exemption)	All SWECS owners	State sales tax of 7-1/2% not paid.	Systems must produce electricity.	Home built systems: submit form and receipts to receive rebate. Manufactured systems not taxed at purchase.		1980-1985
Property Tax Incentive (exemption)	Residential	15 years exemption from assessment for property tax from date of installation in participating towns: 130 out of 169 towns participating.	System must produce electricity to be used at least in part on site.	Owner monitors assessment.		1978 — no expiration date.
Loans	Residential	$400-$3,000 for residential use at 6-1/2%.	Adjusted individual gross income cannot exceed $30,000 per year and individual must live in dwelling of no more than 4 units.	Call toll-free number for application forms.		1979 — no expiration date.
Grants (none)						
Anemometer Loans (none)						

NOTES

Wind Energy Audits Authorized Under Federal RCS Program

CONTACT

Leah Haygood, Wind Program Coordinator
Office Policy and Management
80 Washington Street
Hartford, CT 06115
(203) 566-3394

DELAWARE INCENTIVES FOR SMALL WIND ENERGY CONVERSION SYSTEMS (SWECS)

INCENTIVE	RECIPIENT CATEGORY	PROVISIONS	RESTRICTIONS AND/OR PERFORMANCE CRITERIA	APPLICATION PROCEDURE	CARRYOVER PROVISIONS	DATES OF PROGRAM
Income Tax Incentive (none)						
Sales Tax Incentive (none)						
Property Tax Incentive (none)						
Loans (none)						
Grants	Delaware Residents	Amounts up to $1,000 for residences and $2,500 for farms and businesses to install renewable energy systems. Total funding for program is $100,000.	Being formulated at present time.	Contact State Energy Office		9/80 — until funding depleted.
Anemometer Loan (none)						

NOTES

State hopes to start anemometer loan program next year.
Wind Energy Audits Authorized
Under Federal RCS Program

CONTACT

Dan Anstine, Assistant Director
Renewable Resource Development
Delaware Energy Office
Dover, DE 19901
(302) 736-5644

FLORIDA INCENTIVES FOR SMALL WIND ENERGY CONVERSION SYSTEMS (SWECS)

INCENTIVE	RECIPIENT CATEGORY	PROVISIONS	RESTRICTIONS AND/OR PERFORMANCE CRITERIA	APPLICATION PROCEDURE	CARRYOVER PROVISIONS	DATES OF PROGRAM
Income Tax Incentive (none)						
Sales Tax Incentive (none)	All	No sales tax	Heating only or cooling	Contact state energy office		1980-1990
Property Tax Incentive (none)	Residential	No property tax				1980-1990
Loans (none)						
Grants (none)						
Anemometer Loan (none)						

NOTES
Wind Energy Audits Authorized In Some Counties Under Federal RCS Program

CONTACT

Colleen Kettles
Florida Solar Energy Center
300 State Road 401
Cape Canaveral, FL 32920
(305) 783-0300

GEORGIA INCENTIVES FOR SMALL WIND ENERGY CONVERSION SYSTEMS (SWECS)

INCENTIVE	RECIPIENT CATEGORY	PROVISIONS	RESTRICTIONS AND/OR PERFORMANCE CRITERIA	APPLICATION PROCEDURE	CARRYOVER PROVISIONS	DATES OF PROGRAM
Income Tax Incentive (none)						
Sales Tax Incentive (none)			NO INCENTIVES AS OF MARCH 1982			
Property Tax Incentive (none)						
Loans (none)						
Grants (none)						
Anemometer Loan (none)						

NOTES
Wind Energy Audits Not Required Under Federal RCS Program

CONTACT

Wayne Robertson
Office of Energy Resources
270 Washington Street, S.W.
Atlanta, GA 30334
(404) 656-5176

HAWAII INCENTIVES FOR SMALL WIND ENERGY CONVERSION SYSTEMS (SWECS)

INCENTIVE	RECIPIENT CATEGORY	PROVISIONS	RESTRICTIONS AND/OR PERFORMANCE CRITERIA	APPLICATION PROCEDURE	CARRYOVER PROVISIONS	DATES OF PROGRAM
Income Tax Incentive (none)						
Sales Tax Incentive (none)						
Property Tax Incentive (none)						
Loans (none)						
Grants (none)						
Anemometer Loan	Hawaii Residents.	Loan of testing equipment to monitor wind speed on site for up to 6 months.	None.	Contact Department of Planning.		1979 – no expiration date.

NOTES

Wind Energy Audits Authorized Under Federal RCS Program

CONTACT

Dr. James Berger, Acting Manager
Department of Planning
Box 2359
Honolulu, HI 96804
(808) 548-4195

IDAHO INCENTIVES FOR SMALL WIND ENERGY CONVERSION SYSTEMS (SWECS)

INCENTIVE	RECIPIENT CATEGORY	PROVISIONS	RESTRICTIONS AND/OR PERFORMANCE CRITERIA	APPLICATION PROCEDURE	CARRYOVER PROVISIONS	DATES OF PROGRAM
Income Tax Incentive (deduction)	Residential	Deducation of installed system cost up to $20,000 allowed over a 4-year period – 40% deductible in the first year, not to exceed $5,000. In addition to Federal credit.	None	Included in standard tax forms.	See Provisions.	1978 – no expiration date.
Sales Tax Incentive (none)						
Property Tax Incentive (none)						
Loans (none)						
Grants (none)						
Anemometer Loan (none)						

NOTES

Wind Energy Audits Not Required Under Federal RCS Program

CONTACT

Megan Morgan, Energy Specialist
Idaho Department of Energy
State House Mail
Boise, ID 83720
(208) 334-3800

ILLINOIS INCENTIVES FOR SMALL WIND ENERGY CONVERSION SYSTEMS (SWECS)

INCENTIVE	RECIPIENT CATEGORY	PROVISIONS	RESTRICTIONS AND/OR PERFORMANCE CRITERIA	APPLICATION PROCEDURE	CARRYOVER PROVISIONS	DATES OF PROGRAM
Income Tax Incentive (none)						
Sales Tax Incentive (none)						
Property Tax Incentive (exemption)	All SWECS owners.	Exempts SWECS from assessment for property tax. No time limit.	None yet.	Owner monitors assessment.		1977 — no expiration date.
Loans (none)						
Grants	Illinois residents.	Grants of any amount available from program with original $5,000,000 funding for alternative energy systems.	Must demonstrate economic feasibility and widespread applicability.	Contact Institute of Natural Resources.		1977 — until funding depleted.
Anemometer Loan (none)						

NOTES

State Energy Office estimates 20 SWECS in state.

Approximately 6 wind proposals sent to Grant Board the first year but none funded; board emphasizing gasahol production.

Wind Energy Audits Authorized In Some Counties Under Federal RCS Program

CONTACT

Gary Mielke
Institute of Natural Resources
325 West Adams Street
Springfield, Illinois 62706
(217) 785-0979

INDIANA INCENTIVES FOR SMALL WIND ENERGY CONVERSION SYSTEMS (SWECS)

INCENTIVE	RECIPIENT CATEGORY	PROVISIONS	RESTRICTIONS AND/OR PERFORMANCE CRITERIA	APPLICATION PROCEDURE	CARRYOVER PROVISIONS	DATES OF PROGRAM
Income Tax Incentive (none)						
Sales Tax Incentive (none)						
Property Tax Incentive (exemption)	Residential	Exemption of SWECS from assessment for property tax. No time limit.	Must produce electricity.	Owner monitors assessment.		1978 – no expiration date.
Loans (none)						
Grants (none)						
Anemometer Loan (none)						

NOTES

A tax credit of 25% of installed system cost is being formulated now for 1980 taxes, applicable to all SWECS applications conforming with safety standards being developed.

Wind Energy Audits Authorized In Some Counties Under Federal RCS Program

CONTACT

Robert Joy
Department of Revenue, Room 209
100 North Senate
Indianapolis, IN 46204
(317) 232-2261

IOWA INCENTIVES FOR SMALL WIND ENERGY CONVERSION SYSTEMS (SWECS)

INCENTIVE	RECIPIENT CATEGORY	PROVISIONS	RESTRICTIONS AND/OR PERFORMANCE CRITERIA	APPLICATION PROCEDURE	CARRYOVER PROVISIONS	DATES OF PROGRAM
Income Tax Incentive (none)						
Sales Tax Incentive (none)						
Property Tax Incentive (exemption)	All SWECS owners.	Exemption of SWECS from assessment for property tax until 1985.	None	Owner monitors assessment.		1979-1985
Loans	Residential.	Legislation passed 7/80, details and funding not established.				1980 – Indefinite.
Grants (none)						
Anemometer Loan (none)						

NOTES
- State Energy Office estimates 25-30 SWECS property tax exemptions given.
- Wind Energy Audits Authorized Under Federal RCS Program

CONTACT

Jim Dye
Energy Information Specialist
Iowa Energy Policy Council
217 East 7th Street
Des Moines, Iowa 50319
(515) 281-4420

KANSAS INCENTIVES FOR SMALL WIND ENERGY CONVERSION SYSTEMS (SWECS)

INCENTIVE	RECIPIENT CATEGORY	PROVISIONS	RESTRICTIONS AND/OR PERFORMANCE CRITERIA	APPLICATION PROCEDURE	CARRYOVER PROVISIONS	DATES OF PROGRAM
Income Tax Incentive (credit)	Residential	Credit of 30% of installed system cost up to $1500. In addition to Federal credit.	None	Tax form notes additional form.	3 years. Possible refund for 3 years as well.	1980-1983
Income Tax Incentive (credit)	Commercial Farms, Businesses	Credit of 30% of installed system cost up to $4500. In addition to Federal credit.	None	Noted on tax form.	5 years	1980-1983
Sales Tax Incentive (none)						
Property Tax Incentive (exemption)	All SWECS Owners	Exempts SWECS from assessment for property tax until 1985.	None	Owner monitors assessment.		1980-1985
Loans (none)						
Grants (none)						
Anemometer Loan (none)						

NOTES

State Energy Office estimates at least 60 WECS installed.

Wind Energy Audits Authorized Under Federal RCS Program

CONTACT

David Martin, Coordinator of Solar and Wind Program
Kansas Energy Office
214 West 6th Street
Topeka, KA 66603
(913) 296-2496

KENTUCKY INCENTIVES FOR SMALL WIND ENERGY CONVERSION SYSTEMS (SWECS)

INCENTIVE	RECIPIENT CATEGORY	PROVISIONS	RESTRICTIONS AND/OR PERFORMANCE CRITERIA	APPLICATION PROCEDURE	CARRYOVER PROVISIONS	DATES OF PROGRAM
Income Tax Incentive (none)						
Sales Tax Incentive (none)			NO INCENTIVES AS OF MARCH 1982			
Property Tax Incentive (none)						
Loans (none)						
Grants (none)						
Anemometer Loan (none)						

NOTES

Wind Energy Audits Authorized In Some Counties Under Federal RCS Program

CONTACT

Randal H. Ihara
Office of Planning & Evaluation
Department of Energy
Box 11888
Lexington, KY 40578
(606) 252-5535

LOUISIANA INCENTIVES FOR SMALL WIND ENERGY CONVERSION SYSTEMS (SWECS)

INCENTIVE	RECIPIENT CATEGORY	PROVISIONS	RESTRICTIONS AND/OR PERFORMANCE CRITERIA	APPLICATION PROCEDURE	CARRYOVER PROVISIONS	DATES OF PROGRAM
Income Tax Incentive (none)						
Sales Tax Incentive (none)			NO INCENTIVES AS OF MARCH 1982			
Property Tax Incentive (none)						
Loans (none)						
Grants (none)						
Anemometer Loan (none)						

NOTES
Wind Energy Audits Authorized In Some Counties Under Federal RCS Program

CONTACT

Keith Overdyke
Dept. of Natural Resources and Development
Box 44396
Baton Rouge, LA 70804
(504) 342-4500

MAINE INCENTIVES FOR SMALL WIND ENERGY CONVERSION SYSTEMS (SWECS)

INCENTIVE	RECIPIENT CATEGORY	PROVISIONS	RESTRICTIONS AND/OR PERFORMANCE CRITERIA	APPLICATION PROCEDURE	CARRYOVER PROVISIONS	DATES OF PROGRAM
Income Tax Incentive (credit)	Residential	Credit of 20% of the first $500 spent on system, up to $100 in addition to federal.	System must be purchased after 1/1/79.	Included in standard tax form.	None	1979 — no expiration date.
Sales Tax Incentive (none)						
Property Tax Incentive (none)						
Loans (none)						
Grants (none)						
Anemometer Loan (none)						

NOTES

8 SWECS tax credits granted for 1979.

Wind Energy Audits Authorized Under Federal RCS Program.

CONTACT

Vincent Dicara
Supervisor of Information
Office of Energy Resources
55 Capitol Street
Augusta, ME 04330
(207) 289-3811

MARYLAND INCENTIVES FOR SMALL WIND ENERGY CONVERSION SYSTEMS (SWECS)

INCENTIVE	RECIPIENT CATEGORY	PROVISIONS	RESTRICTIONS AND/OR PERFORMANCE CRITERIA	APPLICATION PROCEDURE	CARRYOVER PROVISIONS	DATES OF PROGRAM
Income Tax Incentive (none)						
Sales Tax Incentive (none)			NO INCENTIVES AS OF MARCH 1982			
Property Tax Incentive (none)						
Loans (none)						
Grants (none)						
Anemometer Loan (none)						

NOTES

Wind Energy Audits Authorized In Some Counties Under Federal RCS Program

CONTACT

Richard Keller
Dept. of Natural Resources
Energy Office – Suite 1302
301 W. Preston Street
Baltimore, MD 21201
(301) 383-6810

MASSACHUSETTS INCENTIVES FOR SMALL WIND ENERGY CONVERSION SYSTEMS (SWECS)

INCENTIVE	RECIPIENT CATEGORY	PROVISIONS	RESTRICTIONS AND/OR PERFORMANCE CRITERIA	APPLICATION PROCEDURE	CARRYOVER PROVISIONS	DATES OF PROGRAM
Income Tax Incentive (credit)	Residential	Credit of 35% of installed system cost after subtracting 40% for Federal credit, up to $1000.00	None	Tax form notes additional form.	3 years	1979-1983
Income Tax Incentive (deduction)	Commercial Farms, Businesses	100% of installed system cost may be deducted from gross income during first year of system operation.	None	Noted on tax form.	None	1977-1983
Sales Tax Incentive (exemption)	Residential	State sales tax of 5% of system cost not paid.	None	Not taxed at purchase.		1977 – no expiration date
Property Tax Incentive (exemption)	Residential	Exempts SWECS from assessment for property tax for 20 years after installation.	None	Owner monitors assessment.		1977 – no expiration date
Loans (none)						
Grants (none)						
Anemometer Loan (none)						

NOTES

Wind Energy Audits Authorized Under Federal RCS Program

CONTACT

George Lagassa, Program Manager for Small Scale Energy Products
State Solar Office
73 Tremont Street
Boston, MA 02108
(617) 727-7297

MICHIGAN INCENTIVES FOR SMALL WIND ENERGY CONVERSION SYSTEMS (SWECS)

INCENTIVE	RECIPIENT CATEGORY	PROVISIONS	RESTRICTIONS AND/OR PERFORMANCE CRITERIA	APPLICATION PROCEDURE	CARRYOVER PROVISIONS	DATES OF PROGRAM
Income Tax Incentive (credit)	Residential	Credit of 25% of first $2,000 and 15% of next $8,000 of installed system cost, up to $1,700. Credit drops each year to 10% and 5% in 1983. In addition to Federal credit.	No installation costs accepted for home-built systems. Manufacturers have state certified systems. Others must submit plans and specifications for approval.	Submit special form, receipts, and proof of home ownership.	None. A full rebate of excess credit amount is given.	1979-1983
Sales Tax Incentive (rebate)	All SWECS Owners	SWECS buyer receives rebate of 4% sales tax paid on system.	None	Obtain special form from state Treasury and submit with receipt.		1976-1985
Property Tax Incentive (exemption)	All SWECS Owners	Perpetual exemption of SWECS from assessment for property tax.	Exemption must be obtained by mid-1985.	Submit special form to locality and state. Home built systems: submit plans in addition.		1976-1985
Loans (none)						
Grants (none)						
Anemometer Loan (none)						

NOTES

20 WECS income tax credits granted for 1979.

Wind Energy Audits Authorized Under Federal RCS Program

CONTACT

Fred Frankena, Information Resources
Michigan Energy Administration
P.O. Box 30228
Lansing, MI 48909
(517) 373-6430

MINNESOTA INCENTIVES FOR SMALL WIND ENERGY CONVERSION SYSTEMS (SWECS)

INCENTIVE	RECIPIENT CATEGORY	PROVISIONS	RESTRICTIONS AND/OR PERFORMANCE CRITERIA	APPLICATION PROCEDURE	CARRYOVER PROVISIONS	DATES OF PROGRAM
Income Tax Incentive (credit)	Residential	Credit of 20% of first $10,000 of installed system cost up to $2,000. In addition to Federal credit although tax paid on Federal credit the following year.	Premises must contain 6 dwelling units or less.	Tax form package includes energy credit form.	Through 1984	1978-1984
Sales Tax Incentive (none)						
Property Tax Incentive (exemption)	Residential	Exempts SWECS from assessment for property tax until 1984.	System must be installed prior to 1984.	Owner monitors assessment.		1978-1984
Loans (none)						
Grants (none)						
Anemometer Loan (none)						

NOTES

State hopes to start anemometer loan program. Two utilities are currently loaning such equipment to their customers.

Wind Energy Audits Authorized
Under Federal RCS Program

CONTACT

John Cutty, Wind Energy Program Manager
160 Kellogg Boulevard
St. Paul, MN 55101
(612) 297-2327

MISSISSIPPI INCENTIVES FOR SMALL WIND ENERGY CONVERSION SYSTEMS (SWECS)

INCENTIVE	RECIPIENT CATEGORY	PROVISIONS	RESTRICTIONS AND/OR PERFORMANCE CRITERIA	APPLICATION PROCEDURE	CARRYOVER PROVISIONS	DATES OF PROGRAM
Income Tax Incentive (none)						
Sales Tax Incentive (none)						
Property Tax Incentive (none)		NO INCENTIVES AS OF MARCH 1982				
Loans (none)						
Grants (none)						
Anemometer Loan (none)						

NOTES
Wind Energy Audits Authorized In Some Counties Under Federal RCS Program

CONTACT

Sheila Molony
Mississippi Office of Energy
Dept. of Natural Resources
Box 10586
Jackson, MS 39205
(601) 961-5099

MISSOURI INCENTIVES FOR SMALL WIND ENERGY CONVERSION SYSTEMS (SWECS)

INCENTIVE	RECIPIENT CATEGORY	PROVISIONS	RESTRICTIONS AND/OR PERFORMANCE CRITERIA	APPLICATION PROCEDURE	CARRYOVER PROVISIONS	DATES OF PROGRAM
Income Tax Incentive (none)						
Sales Tax Incentive (none)						
Property Tax Incentive (none)						
Loans	Missouri Residents	State subsidizes interest on loans for renewable energy systems in various Missouri banks.	Determined by individual banks.	Contact banks.		1979-1980 (extendable)
Grants (none)						
Anemometer Loan (none)						

NOTES

Wind Energy Audits Authorized In Some Counties Under Federal RCS Program

CONTACT

Herb Wade, Solar Manager
Department of Natural Resources
Jefferson City, MO 65101
(317) 751-4000

MONTANA INCENTIVES FOR SMALL WIND ENERGY CONVERSION SYSTEMS (SWECS)

INCENTIVE	RECIPIENT CATEGORY	PROVISIONS	RESTRICTIONS AND/OR PERFORMANCE CRITERIA	APPLICATION PROCEDURE	CARRYOVER PROVISIONS	DATES OF PROGRAM
Income Tax Incentive (credit)	Residential	Credit of 5% of first $1,000 plus 2.5% 2.5% of next $3,000 of installed system cost, up to $125. In addition to Federal credit.	None	Tax form notes additional form.	None	1977 – no expiration date
Sales Tax Incentive (none)						
Property Tax Incentive (exemption)	All SWECS Owners	Exempts SWECS from assessment for property tax for ten years after installation.	Up to $20,000 limit for residences. Up to $100,000 limit for businesses.	Owner monitors assessment.		1979 – no expiration date
Loans (none)						
Grants (Renewable Energy Bureau)	Montana Residents	Grants for projects which demonstrate use of renewable energy.	As decided by Renewable Energy Bureau.	Contact Department of Natural Resources.		1976 – no expiration date.
Anemometer Loan (none)						

NOTES

25-30 grants have been awarded to wind projects.

Wind Energy Audits Not Required Under Federal RCS Program

CONTACT

Tom Livers
Department of Natural Resources
32 South Ewing
Helena, MT 59601
(406) 449-4624

NEBRASKA INCENTIVES FOR SMALL WIND ENERGY CONVERSION SYSTEMS (SWECS)

INCENTIVE	RECIPIENT CATEGORY	PROVISIONS	RESTRICTIONS AND/OR PERFORMANCE CRITERIA	APPLICATION PROCEDURE	CARRYOVER PROVISIONS	DATES OF PROGRAM
Income Tax Incentive (none)						
Sales Tax Incentive (rebate)	All SWECS owners.	SWECS buyer receives rebate of 3 1/2% sales tax paid on system.	System must supplant the use of petroleum.	Submit special form and system plans.		1980-1985
Property Tax Incentive (none)						
Loans (none)						
Grants (none)						
Anemometer Loan (none)						

NOTES

Wind Energy Audits Authorized In Some Counties Under Federal RCS Program

CONTACT

Buck Balok, Deputy Director
Nebraska State Solar Office
W191 Nebraska Hall
Lincoln, NE 68588
(402) 471-2867

NEVADA INCENTIVES FOR SMALL WIND ENERGY CONVERSION SYSTEMS (SWECS)

INCENTIVE	RECIPIENT CATEGORY	PROVISIONS	RESTRICTIONS AND/OR PERFORMANCE CRITERIA	APPLICATION PROCEDURE	CARRYOVER PROVISIONS	DATES OF PROGRAM
Income Tax Incentive (none)						
Sales Tax Incentive (none)						
Property Tax Incentive (exemption)	Residential	Exemption of SWECS from assessment for property tax. No time limit.	SWECS must be used for heating or cooling purposes.	Contact state Energy office for system approval.		1979-1981
Loans (none)						
Grants (none)						
Anemometer Loan (none)						

NOTES

No exemptions granted: No one using a SWECS for heating or cooling.

Wind Energy Audits Authorized In Some Counties Under Federal RCS Program

CONTACT

Bob Loux, Administrator of
Energy and Development
Department of Energy
1050 East Williams St. Suite 405
Carson City, NV 89710
(702) 885-5157

NEW HAMPSHIRE INCENTIVES FOR SMALL WIND ENERGY CONVERSION SYSTEMS (SWECS)

INCENTIVE	RECIPIENT CATEGORY	PROVISIONS	RESTRICTIONS AND/OR PERFORMANCE CRITERIA	APPLICATION PROCEDURE	CARRYOVER PROVISIONS	DATES OF PROGRAM
Income Tax Incentive (none)						
Sales Tax Incentive (none)						
Property Tax Incentive (abatement)	Residential	Local option only. Extent of abatement of tax due to SWECS value determined by local assessors.	Determined by local authorities.	Contact local town offices for eligibility.		1975 — no expiration date
Loans (none)						
Grants (none)						
Anemometer Loan (none)						

NOTES

Wind Energy Audits Authorized In Some Counties Under Federal RCS Program

CONTACT

James McConaha, Renewable Energy Division Director
Solar Energy Office
2 1/2 Beacon Street
Concord, NH 03301
(603) 271-2711

NEW JERSEY INCENTIVES FOR SMALL WIND ENERGY CONVERSION SYSTEMS (SWECS)

INCENTIVE	RECIPIENT CATEGORY	PROVISIONS	RESTRICTIONS AND/OR PERFORMANCE CRITERIA	APPLICATION PROCEDURE	CARRYOVER PROVISIONS	DATES OF PROGRAM
Income Tax Incentive (none)						
Sales Tax Incentive (exemption)	All SWECS owners	State sales tax of 5% of system cost not paid.	System must conform to broad state construction code guidelines.	Not taxed at purchase. Contact construction code office for system eligibility.		1978 – no expiration date.
Property Tax Incentive (exemption)	All SWECS owners	Exemption of SWECS from assessment for property tax until 1982.	Manufactured on kit systems used for heating or cooling of buildings only.	SWECS owner fills out special form prior to construction. Approval is given by local Construction Code official.		1978-1982 (extendable)
Loans (none						
Grants (none)						
Anemometer Loan (none)						

NOTES

Wind Energy Audits Authorized Under Federal RCS Program

CONTACT

Bill Groth, Energy Analyst
Department of Energy
101 Commerce Street
Newark, NJ 07102
(201) 648-6293

NEW MEXICO INCENTIVES FOR SMALL WIND ENERGY CONVERSION SYSTEMS (SWECS)

INCENTIVE	RECIPIENT CATEGORY	PROVISIONS	RESTRICTIONS AND/OR PERFORMANCE CRITERIA	APPLICATION PROCEDURE	CARRYOVER PROVISIONS	DATES OF PROGRAM
Income Tax Incentive	Residential Agricultural Commercial	Not available				
Sales Tax Incentive (none)						
Property Tax Incentive (none)						
Loans (none)						
Grants (none)						
Anemometer Loan (none)						

NOTES

Wind Energy Audits Authorized Under Federal RCS Program

CONTACT

Kenneth M. Barnett
New Mexico Solar Energy Institute
Box 3 SOL
Las Cruces, NM 88003
(505) 646-1846

NEW YORK INCENTIVES FOR SMALL WIND ENERGY CONVERSION SYSTEMS (SWECS)

INCENTIVE	RECIPIENT CATEGORY	PROVISIONS	RESTRICTIONS AND/OR PERFORMANCE CRITERIA	APPLICATION PROCEDURE	CARRYOVER PROVISIONS	DATES OF PROGRAM
Income Tax Incentive (none)						
Sales Tax Incentive (none)						
Property Tax Incentive (exemption)	All SWECS owners	Exempts SWECS from assessment for property tax for 15 years after installation.	System must be installed by 7/1/88.	Submit special forms.		1977-1988
Loans (none)						
Grants (none)						
Anemometer Loan (none)						

NOTES

Wind Energy Audits Authorized In Some Counties Under Federal RCS Program

CONTACT

Paul Wendelgass, Energy Analyst
New York State Energy Office
Agency Building No. 2
Empire State Plaza
Albany, NY 12223
(518) 473-8251

NORTH CAROLINA INCENTIVES FOR SMALL WIND ENERGY CONVERSION SYSTEMS (SWECS)

INCENTIVE	RECIPIENT CATEGORY	PROVISIONS	RESTRICTIONS AND/OR PERFORMANCE CRITERIA	APPLICATION PROCEDURE	CARRYOVER PROVISIONS	DATES OF PROGRAM
Income Tax Incentive	Residential Commercial Agricultural	10% up to $1,000	Generate electricity			No expiration date
Sales Tax Incentive (none)						
Property Tax Incentive (none)						
Loans (none)						
Grants (none)						
Anemometer Loan (none)						

NOTES

Wind Energy Audits Authorized In Some Counties Under Federal RCS Program

CONTACT

John Manuel
Dept. of Commerce
Energy Division
Box 25249
Raleigh, NC 27611
(919) 733-4492

NORTH DAKOTA INCENTIVES FOR SMALL WIND ENERGY CONVERSION SYSTEMS (SWECS)

INCENTIVE	RECIPIENT CATEGORY	PROVISIONS	RESTRICTIONS AND/OR PERFORMANCE CRITERIA	APPLICATION PROCEDURE	CARRYOVER PROVISIONS	DATES OF PROGRAM
Income Tax Incentive (credit)	Residential	Credit of 10% of installed system cost over 2 year period, 5% each year. Federal tax credit is taxed by state as income.	None	Included in standard tax form.	None	1977 – no expiration date
Income Tax Incentive (credit)	Commercial Farms, Businesses	Same as above.	None	Included in standard tax form.	None	1977 – no expiration date
Sales Tax Incentive (exemption)	All SWECS owners	State sales tax of 3% of item not paid.	None	Not taxed at purchase.		1977-1985
Property Tax Incentive (exemption)	All SWECS owners	Exempts SWECS from assessment for property tax for 1 year.	None	Submit special form.		1977-1985
Loans (none)						
Grants (none)						
Anemometer Loan (none)						

NOTES

Wind Energy Audits Authorized Under Federal RCS Program

CONTACT

John Conrad, Energy Specialist
Office of Energy Management and Programs
1533 North 12th Street
Bismarck, ND 58501
(701) 224-2250

OHIO INCENTIVES FOR SMALL WIND ENERGY CONVERSION SYSTEMS (SWECS)

INCENTIVE	RECIPIENT CATEGORY	PROVISIONS	RESTRICTIONS AND/OR PERFORMANCE CRITERIA	APPLICATION PROCEDURE	CARRYOVER PROVISIONS	DATES OF PROGRAM
Income Tax Incentive (credit)	Residential	Credit of 10% of installed system cost up to $1,000. One year delay: system bought in 1979 would obtain credit for 1980 tax year. In addition to Federal credit.	Still formulating. Credit may not exceed tax liability for given year.	Tax form notes additional form.	2 years	1979-1985
Income Tax Incentive (credit)	Commercial Farms, Businesses	Credit of 10% of installed system cost. One year delay (see above). No dollar limit. In addition to Federal credit.	Still formulating. Credit may not exceed tax liability for given year.	Tax form notes additional form.	None	1979-1985
Sales Tax Incentive (exemption)	All SWECS owners	State sales tax of 4% of item cost is rebated.	None yet.	Dealer submits form after sale is made.		
Property Tax Incentive (exemption)	All SWECS owners	Exempts SWECS perpetually from assessment for property tax.	Systems must be installed between 8/79 and 12/85.	Submit special form with copy of building permit.		1979-1985
Loans (none)						
Grants (none)						
Anemometer Loan (none)						

NOTES

Wind Energy Audits Authorized In Some Counties Under Federal RCS Program

CONTACT

Leon Winget, Director of Solar Office
Ohio Department of Energy
30 East Broad Street
Columbus, OH 43215
(614) 466-6797

OKLAHOMA INCENTIVES FOR SMALL WIND ENERGY CONVERSION SYSTEMS (SWECS)

INCENTIVE	RECIPIENT CATEGORY	PROVISIONS	RESTRICTIONS AND/OR PERFORMANCE CRITERIA	APPLICATION PROCEDURE	CARRYOVER PROVISIONS	DATES OF PROGRAM
Income Tax Incentive (credit)	Residential	Credit of 80% of Federal credit on installed system cost of $10,000 or less, dropping 10% each year until expiration in 1989. In addition to Federal credit.	None	Tax form notes additional form. Submit itemized cost of system and site work.	5 years	1980-1989
Income Tax Incentive (credit)	Commercial Farms, Businesses	Credit of 15% of installed system cost. No dollar limit. In addition to Federal credit.	None	Same as above.	5 years	1980-1989
Sales Tax Incentive (none)						
Property Tax Incentive (none)						
Loans (none)						
Grants (none)						
Anemometer Loan (none)						

NOTES

Wind Energy Audits Authorized Under Federal RCS Program

CONTACT

Jim Muehlenweg, Solar Project Office
Department of Energy
4400 North Lincoln Boulevard
Oklahoma City, OK
(405) 521-3941

OREGON INCENTIVES FOR SMALL WIND ENERGY CONVERSION SYSTEMS (SWECS)

INCENTIVE	RECIPIENT CATEGORY	PROVISIONS	RESTRICTIONS AND/OR PERFORMANCE CRITERIA	APPLICATION PROCEDURE	CARRYOVER PROVISIONS	DATES OF PROGRAM
Income Tax Incentive (credit)	Residential	Credit of 25% of installed system cost up to $1,000. In addition to Federal credit.	Area must be anemometer tested. System must provide 10% of household electrical energy needs. System must be warrantied by installer, have five operating machines or be tested at Rocky Flats.	Submit special forms, supporting data, and plans prior to installation.	5 years	1978-1984
Income Tax Incentive (credit)	Commercial Farms, Businesses	Credit of 35% of installed system cost after subtracting the Federal credit. Credit is given over a 5-year period.		Same as above.	None	1980-1984
Sales Tax Incentive (none)						
Property Tax Incentive (exemption)	All SWECS owners	SWECS exempted from assessment for property tax until 1998.				1980-1998
Loans (alternative energy)	Oregon residents	In process of formulation.	See Residential, above.	In process of formulation.		1980 – no expiration date
Grants (none)						
Anemometer Loan	Oregon residents	Loan of wind-speed monitoring equipment for up to 1 year.	None	Contact Energy Office.		1981-1982

NOTES

State energy office estimates 16 electrical SWECS in state, 36 credit applications waiting for approval, and several hundred mechanical systems in state. State loans approximately 35 anemometers with a waiting list of perhaps 200+ people. Utilities will also be loaning anemometer equipment.

Wind Energy Audits Not Required Under Federal RCS Program

CONTACT

Mr. Donald Bain
Oregon Department of Energy
Labor and Industrial Building
Salem, OR 97310
(503) 378-4040

145

PENNSYLVANIA INCENTIVES FOR SMALL WIND ENERGY CONVERSION SYSTEMS (SWECS)

INCENTIVE	RECIPIENT CATEGORY	PROVISIONS	RESTRICTIONS AND/OR PERFORMANCE CRITERIA	APPLICATION PROCEDURE	CARRYOVER PROVISIONS	DATES OF PROGRAM
Income Tax Incentive (none)						
Sales Tax Incentive (none)			NO INCENTIVES AS OF MARCH 1982			
Property Tax Incentive (none)						
Loans (none)						
Grants (none)						
Anemometer Loan (none)						

NOTES

Wind Energy Audits Authorized In Some Counties Under Federal RCS Program

CONTACT

Linda Deliberty
Governors Energy Council
1625 N. Front Street
Harrisburg, PA 17101
(717) 783-8610

146

RHODE ISLAND INCENTIVES FOR SMALL WIND ENERGY CONVERSION SYSTEMS (SWECS)

INCENTIVE	RECIPIENT CATEGORY	PROVISIONS	RESTRICTIONS AND/OR PERFORMANCE CRITERIA	APPLICATION PROCEDURE	CARRYOVER PROVISIONS	DATES OF PROGRAM
Income Tax Incentive (credit)	Residential	Credit of 10% of installed system cost up to $1,000. In addition to Federal credit.	Credit cannot reduce tax liability below $100 yearly.	Tax form package includes energy credit form.	5 years	1980-1985
Income Tax Incentive (credit)	Commercial Farms, Businesses	Credit of 10% of installed system cost, up to $1,500. In addition to Federal credit.	System must generate electricity or provide heating or cooling. Credit cannot reduce tax liability below $100 yearly.	Tax form package includes energy credit form.	5 years	1980-1985
Sales Tax Incentive (exemption)	All SWECS owners	State sales tax of 6% not paid.	None	Not taxed at purchase.		1980-1985
Property Tax Incentive (exemption)	All SWECS owners	SWECS cannot be assessed at more than conventional system serving same function.	None	Owner monitors assessment.		1980 – no expiration date
Loans (none)						
Grants (none)						
Anemometer Loan (none)						

NOTES

Wind Energy Audits Authorized Under Federal RCS Program

CONTACT

Bob Ericson
Renewable Energy Planner
Governor's Energy Office
80 Dean Street
Providence, RI 02903

SOUTH CAROLINA INCENTIVES FOR SMALL WIND ENERGY CONVERSION SYSTEMS (SWECS)

INCENTIVE	RECIPIENT CATEGORY	PROVISIONS	RESTRICTIONS AND/OR PERFORMANCE CRITERIA	APPLICATION PROCEDURE	CARRYOVER PROVISIONS	DATES OF PROGRAM
Income Tax Incentive (credit)	All SWECS owners	25% of installed system cost may be deducted from taxable income, while retaining original tax rate. In addition to Federal credit.	In process of formulation.	Tax form notes additional form.	5 years	1980-1983
Sales Tax Incentive (none)						
Property Tax Incentive (none)						
Loans (none)						
Grants (none)						
Anemometer Loan (none)						

NOTES

Wind Energy Audits Authorized In Some Counties Under Federal RCS Program

CONTACT

Roy Shive
Solar Energy Program Manager
Division of Energy Resources
1122 Lady Street
Columbia, SC 29201
(803) 758-8110

SOUTH DAKOTA INCENTIVES FOR SMALL WIND ENERGY CONVERSION SYSTEMS (SWECS)

INCENTIVE	RECIPIENT CATEGORY	PROVISIONS	RESTRICTIONS AND/OR PERFORMANCE CRITERIA	APPLICATION PROCEDURE	CARRYOVER PROVISIONS	DATES OF PROGRAM
Income Tax Incentive (none)						
Sales Tax Incentive (none)						
Property Tax Incentive (assessment credit)	All SWECS owners	Amount of credit declines over 6 year period.	None	Submit special form between November 1st and December 10th of the year.		1980 – no expiration date
Loans (none)						
Grants (none)						
Anemometer Loan (none)						

NOTES
Wind Energy Audits Authorized In Some Countries Under Federal RCS Program

CONTACT

Steve Wagman, Energy Specialist
Office of Energy Policy
Capitol Lake Plaza
Pierre, SD 57501
(605) 773-3603

TENNESSEE INCENTIVES FOR SMALL WIND ENERGY CONVERSION SYSTEMS (SWECS)

INCENTIVE	RECIPIENT CATEGORY	PROVISIONS	RESTRICTIONS AND/OR PERFORMANCE CRITERIA	APPLICATION PROCEDURE	CARRYOVER PROVISIONS	DATES OF PROGRAM
Income Tax Incentive (none)						
Sales Tax Incentive (none)						
Property Tax Incentive (exemption)	All SWECS owners	SWECS exempted from assessment for property tax each year until 1988.	None	Make claim to local tax assessor annually.		1978-1988
Loans (none)						
Grants (none)						
Anemometer Loan (none)						

NOTES

Wind Energy Audits Not Required Under Federal RCS Program

CONTACT

Eugene Edmunds, Manager
Tennessee Energy Authority
Suite 707
Capitol Boulevard
Nashville, TN 37219
(615) 741-2994

TEXAS INCENTIVES FOR SMALL WIND ENERGY CONVERSION SYSTEMS (SWECS)

INCENTIVE	RECIPIENT CATEGORY	PROVISIONS	RESTRICTIONS AND/OR PERFORMANCE CRITERIA	APPLICATION PROCEDURE	CARRYOVER PROVISIONS	DATES OF PROGRAM
Income Tax Incentive (none – no income tax in this state)						
Sales Tax Incentive (exemption)	All SWECS owners	State sales tax of 4% not paid. Builders of their own systems do not pay tax on parts.	None	Not taxed at purchase. Builders of their own systems obtain exemption certificate from sales tax office.		1978 – no expiration date
Property Tax Incentive (exemption)	All SWECS owners	Exempts SWECS from assessment for property tax. No time limit.	None	Owner monitors assessment.		1978 – no expiration date.
Loans (none)						
Grants (none)						
Anemometer Loan (none)						

NOTES

Wind Energy Audits Authorized In Some Counties Under Federal RCS Program

CONTACT

Bob Avant, Coordinator of Biomass and Wind
Texas Energy Advisory Council
411 W. 13th Street
Austin, TX 78701
(512) 475-5588

UTAH INCENTIVES FOR SMALL WIND ENERGY CONVERSION SYSTEMS (SWECS)

INCENTIVE	RECIPIENT CATEGORY	PROVISIONS	RESTRICTIONS AND/OR PERFORMANCE CRITERIA	APPLICATION PROCEDURE	CARRYOVER PROVISIONS	DATES OF PROGRAM
Income Tax Incentive (credit)	Residential	Credit of 10% of installed system cost up to $1,000. Retroactive to 1977. In addition to Federal credit.	Owner must submit plans and summary of performance. Applies to systems installed between 7/1/77 and 7/1/85.	Still being formulated.	4 years	1980-1985 Retroactive to 1977.
Income Tax Incentive (credit)	Commercial Farms, Businesses	Credit of 10% of installed system cost up to $3,000. Retroactive to 1977. In addition to Federal credit.	Same as above.	Still being formulated.	4 years	1980-1985
Sales Tax Incentive (none)						
Property Tax Incentive (none)						
Loans (none)						
Grants (none)						
Anemometer Loan (none)						

NOTES

Wind Energy Audits Not Required Under Federal RCS Program

CONTACT

Kevin Gillars
Renewable Resource Specialist
State Energy Office
321 East 400 South
Salt Lake City, UT
(801) 533-5424

VERMONT INCENTIVES FOR SMALL WIND ENERGY CONVERSION SYSTEMS (SWECS)

INCENTIVE	RECIPIENT CATEGORY	PROVISIONS	RESTRICTIONS AND/OR PERFORMANCE CRITERIA	APPLICATION PROCEDURE	CARRYOVER PROVISIONS	DATES OF PROGRAM
Income Tax Incentive (credit)	Residential	Credit of 25% of the first $4,000 of installed system cost up to $1,000. In addition to Federal credit.	Home-built systems cost cannot include labor. Applies to systems installed 1/1/78 to 7/1/83.	Tax form notes additional form. Home-built systems submit receipts and plans as well.	4 years	1978-1983
Income Tax Incentive (credit)	Commercial farms, businesses	Credit of 25% of the first $1,200 of installed system cost up to $3,000. In addition to Federal credit.	Same as above.	Same as above.	4 years	1978-1983
Sales Tax Incentive (none)						
Property Tax Incentive (exemption)	All SWECS owners	SWECS and surrounding land, not to exceed 1 acre, exempt from assessment for property tax by state enabling legislation. 30 towns participating. No time limit.	No part of system may be for sale to public.	Contact town offices for eligibility.		1976 – no expiration date.
Loans (none)						
Grants (none)						
Anemometer Loans (none)						

NOTES

14 wind credits granted in 1978.
9 wind credits granted in 1979.

Wind Energy Audits Authorized In Some Counties Under Federal RCS Program

CONTACT

Katrine Goska, Energy Information Librarian
Vermont Energy Office
State Office Building
Montpelier, VT 05602
(802) 828-2393

VIRGINIA INCENTIVES FOR SMALL WIND ENERGY CONVERSION SYSTEMS (SWECS)

INCENTIVE	RECIPIENT CATEGORY	PROVISIONS	RESTRICTIONS AND/OR PERFORMANCE CRITERIA	APPLICATION PROCEDURE	CARRYOVER PROVISIONS	DATES OF PROGRAM
Income Tax Incentive (none)						
Sales Tax Incentive (none)						
Property Tax Incentive (exemption)	All SWECS owners	SWECS exempted from assessment for property tax by state enabling legislation. 6 towns participating. Exemption for no less than 5 years.	Office of Uniform Building Code must certify system eligibility at request of local assessors.	Submit special form.		1977 – no expiration date.
Loans (none)						
Grants (none)						
Anemometer Loan (none)						

NOTES

Wind Energy Audits Authorized In Some Counties Under Federal RCS Program

CONTACT

Temple Bayliff
Director of Energy
Office of Energy
310 Turner Road
Richmond, VA
(804) 745-3245

WASHINGTON INCENTIVES FOR SMALL WIND ENERGY CONVERSION SYSTEMS (SWECS)

INCENTIVE	RECIPIENT CATEGORY	PROVISIONS	RESTRICTIONS AND/OR PERFORMANCE CRITERIA	APPLICATION PROCEDURE	CARRYOVER PROVISIONS	DATES OF PROGRAM
Income Tax Incentive (none)						
Sales Tax Incentive (none)			NO INCENTIVES AS OF MARCH 1982			
Property Tax Incentive (none)						
Loans (none)						
Grants (none)						
Anemometer Loan (none)						

NOTES

Wind Energy Audits Not Required Under Federal RCS Program

CONTACT

Mary S. Anderson
Washington State Energy Office
100 East Union
Olympia, WA 98504
(206) 754-1350

WEST VIRGINIA INCENTIVES FOR SMALL WIND ENERGY CONVERSION SYSTEMS (SWECS)

INCENTIVE	RECIPIENT CATEGORY	PROVISIONS	RESTRICTIONS AND/OR PERFORMANCE CRITERIA	APPLICATION PROCEDURE	CARRYOVER PROVISIONS	DATES OF PROGRAM
Income Tax Incentive (none)						
Sales Tax Incentive (none)			NO INCENTIVES AS OF MARCH 1982			
Property Tax Incentive (none)						
Loans (none)						
Grants (none)						
Anemometer Loan (none)						

NOTES
Wind Energy Audits Not Required Under Federal RCS Program

CONTACT

Charles Coffman
Governors Office of Economic and
Community Development
126 1/2 Greenbrier Street
Charleston, WV 25311
(304) 348-8860

WISCONSIN INCENTIVES FOR SMALL WIND ENERGY CONVERSION SYSTEMS (SWECS)

INCENTIVE	RECIPIENT CATEGORY	PROVISIONS	RESTRICTIONS AND/OR PERFORMANCE CRITERIA	APPLICATION PROCEDURE	CARRYOVER PROVISIONS	DATES OF PROGRAM
Income Tax Incentive (none)						
Sales Tax Incentive (none)						
Property Tax Incentive (none)		NO INCENTIVES AS OF MARCH 1982				
Loans (none)						
Grants	Individuals and Businesses	10% rebate up to $750, for individuals and up to $1000, for businesses		Submit special form		Expires 31 Dec. 1985
Anemometer Loan (none)						

NOTES

Wind Energy Audits Not Required Under Federal RCS Program

CONTACT

Mr. Phil Tyson
State Energy Office
Madison, Wis. 53702
(608) 266-8234

157

WYOMING INCENTIVES FOR SMALL WIND ENERGY CONVERSION SYSTEMS (SWECS)

INCENTIVE	RECIPIENT CATEGORY	PROVISIONS	RESTRICTIONS AND/OR PERFORMANCE CRITERIA	APPLICATION PROCEDURE	CARRYOVER PROVISIONS	DATES OF PROGRAM
Income Tax Incentive (none)						
Sales Tax Incentive (none)						
Property Tax Incentive (none)		NO INCENTIVES AS OF MARCH 1982				
Loans (none)						
Grants (none)						
Anemometer Loan (none)						

NOTES

Wind Energy Audits Authorized In Some Countries Under Federal RCS Program

CONTACT

Ed Maycumber
Capitol Hill Office Bldg.
Energy Conservation Office
25th and Pioneer
Cheyenne, WY 82002
(307) 777-7131

PRESENT VALUE OF ONE DOLLAR DUE AT END OF N YEARS

Present Value of One Dollar Due at the End of n Years

n	1%	2%	3%	4%	5%	6%	7%	8%	9%	10%	n
01	.99010	.98039	.97007	.96154	.95238	.94340	.93458	.92593	.91743	.90909	01
02	.98030	.96117	.94260	.92456	.90703	.89000	.87344	.85734	.84168	.82645	02
03	.97059	.94232	.91514	.88900	.86384	.83962	.81630	.79383	.77218	.75131	03
04	.96098	.92385	.88849	.85480	.82270	.79209	.76290	.73503	.70843	.68301	04
05	.95147	.90573	.86261	.82193	.78353	.74726	.71299	.68058	.64993	.62092	05
06	.94204	.88797	.83748	.79031	.74622	.70496	.66634	.63017	.59627	.56447	06
07	.93272	.87056	.81309	.75992	.71068	.66506	.62275	.58349	.54703	.51316	07
08	.92348	.85349	.78941	.73069	.67684	.62741	.58201	.54027	.50187	.46651	08
09	.91434	.83675	.76642	.70259	.64461	.59190	.54393	.50025	.46043	.42410	09
10	.90529	.82035	.74409	.67556	.61391	.55839	.50835	.46319	.42241	.38554	10
11	.89632	.80426	.72242	.64958	.58468	.52679	.47509	.42888	.38753	.35049	11
12	.88745	.78849	.70138	.62460	.55684	.49697	.44401	.39711	.35553	.31683	12
13	.87866	.77303	.68095	.60057	.53032	.46884	.41496	.36770	.32618	.28966	13
14	.86996	.75787	.66112	.57747	.50507	.44230	.38782	.34046	.29925	.26333	14
15	.86135	.74301	.64186	.55526	.48102	.41726	.36245	.31524	.27454	.23939	15
16	.85282	.72845	.62317	.53391	.45811	.39365	.33873	.29189	.25187	.21763	16
17	.84438	.71416	.60502	.51337	.43630	.37136	.31657	.27027	.23107	.19784	17
18	.83602	.70016	.58739	.49363	.41552	.35034	.29586	.25025	.21199	.17986	18
19	.82774	.68643	.57029	.47464	.39573	.33051	.27651	.23171	.19449	.16351	19
20	.81954	.67297	.55367	.45639	.37689	.31180	.25842	.21455	.17843	.14864	20
21	.81143	.65978	.53755	.43883	.35894	.29415	.24151	.19866	.16370	.13513	21
22	.80340	.64684	.52189	.42195	.34185	.27750	.22571	.18394	.15018	.12285	22
23	.79544	.63414	.50669	.40573	.32557	.26180	.21095	.17031	.13778	.11168	23
24	.78757	.62172	.49193	.39012	.31007	.24698	.19715	.15770	.12640	.10153	24
25	.77977	.60953	.47760	.37512	.29530	.23300	.18425	.14602	.11597	.09230	25

n	11%	12%	13%	14%	15%	16%	17%	18%	19%	20%	n
01	.90090	.89286	.88496	.87719	.86957	.86207	.85470	.84746	.84034	.83333	01
02	.81162	.79719	.78315	.76947	.75614	.74316	.73051	.71818	.70616	.69444	02
03	.73119	.71178	.69305	.67497	.65752	.64066	.62437	.60863	.59342	.57870	03
04	.65873	.63552	.61332	.59208	.57175	.55229	.53365	.51579	.49867	.48225	04
05	.59345	.56743	.54276	.51937	.49718	.47611	.45611	.43711	.41905	.40188	05
06	.53464	.50663	.48032	.45559	.43233	.41044	.38984	.37043	.35214	.33490	06
07	.48166	.45235	.42506	.39964	.37594	.35383	.33320	.31392	.29592	.27908	07
08	.43393	.40388	.37616	.35056	.32690	.30503	.28478	.26604	.24867	.23257	08
09	.39092	.36061	.33288	.30751	.28426	.26295	.24340	.22546	.20897	.19381	09
10	.35218	.32197	.29459	.26974	.24718	.22668	.20804	.19106	.17560	.16151	10
11	.31728	.28748	.26070	.23662	.21494	.19542	.17781	.16192	.14756	.13459	11
12	.28584	.25667	.23071	.20756	.18691	.16846	.15197	.13722	.12400	.11216	12
13	.25751	.22917	.20416	.18207	.16253	.14523	.12989	.11629	.10420	.09346	13
14	.23199	.20462	.18068	.15971	.14133	.12520	.11102	.09855	.08757	.07789	14
15	.20900	.18270	.15989	.14010	.12289	.10793	.09489	.08352	.07359	.06491	15
16	.18829	.16312	.14150	.12289	.10686	.09304	.08110	.07078	.06184	.05409	16
17	.16963	.14564	.12522	.10780	.09293	.08021	.06932	.05998	.05196	.04507	17
18	.15282	.13004	.11081	.09456	.08080	.06914	.05925	.05083	.04367	.03756	18
19	.13768	.11611	.09806	.08295	.07026	.05961	.05064	.04308	.03669	.03130	19
20	.12403	.10367	.08678	.07276	.06110	.05139	.04328	.03651	.03084	.02608	20
21	.11174	.09256	.07680	.06383	.05313	.04430	.03699	.03094	.02591	.02174	21
22	.10067	.08264	.06796	.05599	.04620	.03819	.03162	.02622	.02178	.01811	22
23	.09069	.07379	.06014	.04911	.04017	.03292	.02702	.02222	.01830	.01509	23
24	.08170	.06588	.05322	.04308	.03493	.02838	.02310	.01883	.01538	.01258	24
25	.07361	.05882	.04710	.03779	.03038	.02447	.01974	.01596	.01292	.01048	25

n	21%	22%	23%	24%	25%	26%	27%	28%	29%	30%
01	.82645	.81967	.81301	.80645	.80000	.79365	.78740	.78125	.77519	.76923
02	.68301	.67186	.66098	.65036	.64000	.62988	.62000	.61035	.60093	.59172
03	.56447	.55071	.53738	.52449	.51200	.49991	.48819	.47684	.46583	.45517
04	.46651	.45140	.43690	.42297	.40960	.39675	.38440	.37253	.36111	.35013
05	.38554	.37000	.35520	.34111	.32768	.31488	.30268	.29104	.27993	.26933
06	.31863	.30328	.28878	.27509	.26214	.24991	.23833	.22737	.21700	.20718
07	.26333	.24859	.23478	.22184	.20972	.19834	.18766	.17764	.16822	.15937
08	.21763	.20376	.19088	.17891	.16777	.15741	.14776	.13878	.13040	.12259
09	.17986	.16702	.15519	.14428	.13422	.12493	.11635	.10842	.10109	.09430
10	.14864	.13690	.12617	.11635	.10737	.09915	.09161	.08470	.07836	.07254
11	.12285	.11221	.10258	.09383	.08590	.07869	.07214	.06617	.06075	.05580
12	.10153	.09198	.08339	.07567	.06872	.06245	.05680	.05170	.04709	.04292
13	.08391	.07539	.06780	.06103	.05498	.04957	.04472	.04039	.03650	.03302
14	.06934	.06180	.05512	.04921	.04398	.03934	.03522	.03155	.02830	.02540
15	.05731	.05065	.04481	.03969	.03518	.03122	.02773	.02465	.02194	.01954
16	.04736	.04152	.03643	.03201	.02815	.02478	.02183	.01926	.01700	.01503
17	.03914	.03403	.02962	.02581	.02252	.01967	.01719	.01505	.01318	.01156
18	.03235	.02789	.02408	.02082	.01801	.01561	.01354	.01175	.01022	.00889
19	.02673	.02286	.01958	.01679	.01441	.01239	.01066	.00918	.00792	.00684
20	.02209	.01874	.01592	.01354	.01153	.00983	.00839	.00717	.00614	.00526
21	.01826	.01536	.01294	.01092	.00922	.00780	.00661	.00561	.00476	.00405
22	.01509	.01259	.01052	.00880	.00738	.00619	.00520	.00438	.00369	.00311
23	.01247	.01032	.00855	.00710	.00590	.00491	.00410	.00342	.00286	.00239
24	.01031	.00846	.00695	.00573	.00472	.00390	.00323	.00267	.00222	.00184
25	.00852	.00693	.00565	.00462	.00378	.00310	.00254	.00209	.00172	.00142

n	31%	32%	33%	34%	35%	36%	37%	38%	39%	40%
01	.76336	.75758	.75188	.74627	.74074	.73529	.72993	.72464	.71942	.71429
02	.58272	.57392	.56532	.55692	.54870	.54066	.53279	.52510	.51757	.51020
03	.44482	.43479	.42505	.41561	.40644	.39754	.38890	.38051	.37235	.36443
04	.33956	.32939	.31959	.31016	.30107	.29231	.28387	.27573	.26788	.26031
05	.25920	.24953	.24029	.23146	.22301	.21493	.20720	.19980	.19272	.18593
06	.19787	.18904	.18067	.17273	.17520	.15804	.15124	.14479	.13865	.13281
07	.15104	.14321	.13584	.12890	.12237	.11621	.11040	.10492	.09975	.09486
08	.11530	.10849	.10214	.09620	.09064	.08545	.08058	.07603	.07176	.16776
09	.08802	.08219	.07680	.07179	.06714	.06283	.05882	.05509	.05163	.04840
10	.06719	.06227	.05774	.05357	.04973	.04620	.04293	.03992	.03714	.03457
11	.05129	.04717	.04341	.03998	.03684	.03397	.03134	.02893	.02672	.02469
12	.03915	.03573	.03264	.02984	.02729	.02498	.02287	.02096	.01922	.01764
13	.02989	.02707	.02454	.02227	.02021	.01837	.01670	.01519	.01383	.01260
14	.02281	.02051	.01845	.01662	.01497	.01350	.01219	.01101	.00995	.00900
15	.01742	.01554	.01387	.01240	.01109	.00993	.00890	.00798	.00716	.00643
16	.01329	.01177	.01043	.00925	.00822	.00730	.00649	.00578	.00515	.00459
17	.01015	.00892	.00784	.00691	.00609	.00537	.00474	.00419	.00370	.00328
18	.00775	.00676	.00590	.00515	.00451	.00395	.00346	.00304	.00267	.00234
19	.00501	.00512	.00443	.00385	.00334	.00290	.00253	.00220	.00192	.00167
20	.00451	.00388	.00333	.00287	.00247	.00213	.00184	.00159	.00138	.00120
21	.00345	.00294	.00251	.00214	.00183	.00157	.00135	.00115	.00099	.00085
22	.00263	.00223	.00188	.00160	.00136	.00115	.00098	.00084	.00071	.00061
23	.00201	.00169	.00142	.00119	.00101	.00085	.00072	.00061	.00051	.00044
24	.00153	.00128	.00107	.00089	.00074	.00062	.00052	.00044	.00037	.00031
25	.00117	.00097	.00080	.00066	.00055	.00046	.00038	.00032	.00027	.00022

TECHNICAL WIND ENERGY BIBLIOGRAPHY

This bibliography is a partial listing of reports generated by the Federal Wind Energy Program. Due to the swiftly increasing number of titles dealing with wind energy, this list is restricted to documents produced for or by the federal government in its' research efforts.

These reports are available from:

> National Technical Information Service
> U.S. Department of Commerce
> 5285 Port Royal Road
> Springfield, VA 22161
>
> (703) 557-4650

NTIS will also provide price information. When ordering, the document number, which follows the publication date in each entry, should be specified. Nongovernment books on wind power can often be found in local libraries and bookstores.

CONTENTS

topic
Mission Analyses
Workshops
Applications Studies
Farm Applications
Technology Development (small machines)
Technology Development (large machines)
Vertical-Axis Wind Turbines
Innovative Systems
Wind Characteristics
Legal, Social, Environmental Issues

Mission Analyses

Draper. *Mission Analysis Elements of the Federal Wind Energy Program*. May 15, 1979. AD-03-03-01-1.

General Electric. *Wind Energy Mission Analysis*. Executive Summary, Final Report, Appendices. February 18, 1977. COO/2578-1/1, 2, 3

Lockheed. *Wind Energy Mission Analysis*. Executive Summary, Final Report, Appendix. October 1976. CR 27611.

Rockwell International. *FY 1979 Annual Report. Technical and Management Support for Small Wind Systems*.

Rockwell International. *FY 1980 Program Summary. Technical and Management Support for the Development of Small Wind Systems*. March 1, 1980. REP-3121/3533/80/8.

Rockwell International. *Rocky Flats Small Wind Systems Test Center Activities. Atmospheric Test Data Collected from Small Wind Energy Conversion Systems.* July 1979. RFP-3004. 2 Vols.

SERI. *Basic and Applied Research Program Semiannual Report (July-December 1978).* December 1979. SERI/TR-334-244.

Workshops

Electric Power Research Institute. *Proceedings of the Workshop on Economic and Operational Requirements and Status of Large Scale Wind Systems.* July 1979. EPRI ER-1110-DR.

NSF. *Wind Energy Conversion Systems. Workshop Proceedings.* December 1973. NSF/RA/W-73-006.

NSF. *Proceedings of the Second Workshop on Wind Energy Conversion Systems.* Washington, D.C., June 9-11, 1975. NSF/RA-N-75-050.

JBF. *Third Wind Energy Workshop.* Vol. 1, Vol. 2. May 1978. CONF-770921.

JBF. *Fourth Wind Energy Workshop.* October 1979, Washington, D.C. CONF 791097.

JBF. *Proceedings of the Workshop on Economic and Operational Requirements and Status of Large Scale Wind Systems, March 26-30, 1979.* June 1979. CONF-790352.

NASA. *Large Wind Turbine Design Characteristics and R&D Requirements.* NASA Conference Publication 2106. DOE Publication CONF 790-4111.

NASA. *Wind Energy Conversion Systems.* A workshop sponsored by the NSF-RANN, and NASA, and held in Washington, D.C., June 11-13, 1973. December 1973. DOE/NASA/1010-77/1.

NASA. *Wind Turbine Dynamics.* Proceedings for a workshop held at Cleveland State University, Cleveland, Ohio. February 24-26, 1981. SERI/CP-635-1238. NASA Conference Publication 2185. DOE Publication CONF-810226.

Rockwell International. *Proceedings Small Wind Turbine Systems 1979; A Workshop on R&D Requirements and Utility Interface/Institutional Issues.* Boulder, Colorado. RFP/3014/3533/79-8.

Sandia. *Proceedings of the Vertical Axis Wind Turbine (VAWT) Design Technology Seminar for Industry.* April 1-3, 1980, Albuquerque, New Mexico. August 1980. SAND80-098.

Sandia *Proceedings of the Workshop on Mechanical Storage of Wind Energy.* January 1979. SAND79-0001.

SERI. *Proceedings: Panel on Information Dissemination for Wind Energy August 2nd & 3rd 1979.* Albuquerque, New Mexico. April 1980. SERI/TP-732-343.

SERI. *SERI Second Wind Energy Innovative Systems Conference.* December 3-5, 1980, Colorado Springs, Colorado. SERI/CP-635-938.

Applications Studies

AAI Corporation and Institute of Gas Technology. *Production of Methane Using Offshore Wind Energy.* Final Report. November 1975. Executive Summary.

Aerospace Corporation. *Electric Utility Application of Wind Energy Conversion Systems on the Island of Oahu.* February 23, 1979. ATR-78(7598). Executive Summary and Vol. 2.

Aerospace Corporation. *Wind Machines for the California Aqueduct.* Vols. 1 & 2. March 1977.

Agriculture, U.S. Department of. *On-Farm U.S. Irrigation Pumping Plants.* Final Report. April 30, 1980. DOE/SEA-7315-20741/81/1. Prepared under Interagency Agreement: EX-76-A29-1026.

Argonne National Laboratory. *Reliability, Energy, and Cost Effects of Wind-Powered Generation Integrated with a Conventional Generating System.* January 1980. ANL/AA-17.

Brookhaven National Laboratory. *Wind Power and Electric Utilities: A Review of the Problems and Prospects.* April 1978. BNL50849.

Development Planning and Research Associates, Inc. *Wind Energy Applications in Agriculture.* August 1979. DOE/SEA-1109-20401/79/2.

Electric Power Research Institute. *Large Wind Turbine Generator Performance Assessment. Technology Status Report #3.* July 1981. EPRI AP-1959. Prepared by Arthur D. Little.

Electric Power Research Institute. *Wind Power Generation Dynamic Impacts on Electric Utility Systems.* November 1980. EPRI AP-1614 TPS 79-775.

Honeywell. *The Application of Wind Power Systems to the Service Area of the Minnesota Power & Light Company.* Final Report. Period July 1975-August 1976. October 1977. COO/2618-76-1. Plus Executive Summary COO-2618-1.

JBF. *Wind Energy Systems Application to Regional Utilities.* Executive Summary. Vol. 1, Vol. 2. June 1979.

Massachusetts, University of. *Investigation of the Feasibility of Using Wind-power for Space Heating in Colder Climates.* 1976. NSF/RANN/AER-75-00603/PR/76/1.

Michigan State University. *Application Study of Wind Power Technology to the City of Hart, Michigan, 1977.* January 1978. COO-2992-78/1.

Michigan State University. *Planning Manual for Utility Application of WECS.* June 1979. COO/4450-79/1.

NSF. *Wind Powered Aeration for Remote Locations.* Final Report. 5/15/75-8/31/76. October 1976. ERDA/NSF/-00833/75/1. Grant #AER75-00833.

Rockwell International. *Investigation of the Feasibility of Using Wind Power for Space Heating in Colder Climates.* Annual Report for the Period Ending June 30, 1978. October 1978. DOE/DP-03533-T3.

Rockwell International. *Operating of Small Wind Turbines on a Distribution System.* Final Report March 1981, RFP-3177-2. Prepared by Systems Control, Inc. for Rockwell International under Subcontract #PF-944452.

Rockwell International. *A Preliminary Investigation of Three Advanced Wind Furnace Systems for Residential and Farm Applications.* April 1979. RFP-3059/67025/3533/80/4-1 & 4-2. (2 Vols.-Executive Summary and Technical Report).

Rockwell International. *Small Wind Systems Applications Analysis.* June 1981. RFP-3147/2.

Rockwell International. *SWECS Qualifications for State Programs. Final Report.* July 1980. RFP-3127/05480/80-11.

Rockwell International. *Utility Concerns About Interconnected Small Wind Energy Conversion Systems.* November 1980. TM-IP-8102.

Taywood Engineering Ltd. *Report on Assessment of Offshore Siting of Wind Turbine Generators.* 2 Vols. December 1979. 014D/DED/79/2156.

Westinghouse. *Design Study and Economic Assessment of Multi-Unit Offshore Wind Energy Conversion Systems Application.* Vols. 1, 2, 4. June 14, 1979. WASH-2330-78/4.

Farm Applications

Agriculture, U.S. Department of. *Economic Analysis of Wind-Powered Crop Drying.* Final Report. March 1980. DOE/SEA-1109-20401/81/2. Prepared under Interagency Agreement #: EX-76-A-1026.

Agriculture, U.S. Department of. *Economics of Wind Energy for Irrigation Pumping.* July 14, 1980. DOE/SEA-7315-20741/81/2. Prepared under Interagency Agreement #: EX-76-A-29-1026.

Agriculture, U.S. Department of. *Economic Analysis of Wind-Powered Refrigeration Cooling/Water Heating Systems in Food Processing.* Final Report. March 1980. DOE/SEA-1109-20401/81/1. Prepared under Interagency Agreement: EX-76-A-29-1026.

Agriculture, U.S. Department of. *On-Farm U.S. Irrigation Pumping Plants.* Final Report. April 30, 1980. DOE/SEA-7315-20741/81/1. Prepared under Interagency Agreement: EX-76-A29-1026.

Development Planning and Research Associates, Inc. *Wind Energy Applications in Agriculture.* August 1979. DOE/SEA-1109-20401/79/2.

Kaman Aerospace. *Wind Powered Heat Pump in a Dairy Farm Application.* October 16, 1978. DOE/SEA-1109-20401/79/1.

Kansas State University. *Application of Wind Energy to Great Plains Irrigation Pumping.* October 1979. DOE/SEA-3707-20741/80/1.

Technology Development—Small Machines

Grumman. *Development of an 8 Kilowatt Wind Turbine Generator for Residential Type Applications.* Phase 1. Design and Analysis. Vol. 1, Executive Summary. March 1980. RFP-3007 (Vol. 1). Contract #: AC04-76DP03533 and Vol. 11-Technical Report RFP-3007 (Vol. 2).

Rockwell International. *Altos-Model 8B Wind Turbine Generator. Performance Report.* July 1979. RFP-3033/3533/79-4.

Rockwell International. *Controlled Velocity Testing of Small Wind Energy Conversion Systems. An Evaluation of a Technique.* November 1980. RFP-3189.

Rockwell International. *Development of a 1 kW High-Reliability Wind Turbine Generator.* RFP #PF64410F. May 3, 1977.

Rockwell International. *Development of a 2 Kilowatt High Reliability Wind Machine.* January 1980. RFP-3025/64410/3533/79/16-2.

Rockwell International. *Development of an 8 kW Wind Turbine Generator for Residential Type Applications.* Phase I—Design & Analysis. 2 Vols. June 25, 1979. DOE/DP-103533-T1, Vol. 1-2.

Rockwell International. *Development of a 40 Kilowatt Wind Turbine Generator Phase 1, Design and Analysis (and Executive Summary).* 2 Vols. Prepared for Rockwell International by Kaman Aerospace Corporation. February 1981. RFP-3094-1.

Rockwell International. *Dunlite Model 81-002550 Wind Turbine Generator.* Final Test Report. February 1980. RFP-3149/3533/80-17.

Rockwell International. *Grumman Windstream 25 Wind Turbine Generator. Final Test Report.* March 1980. RFP-3134.

Rockwell International. *Millville Wind Turbine Generator Failure Analysis and Corrective Design Modification.* July 1979. RFP-2992/3533/79-3.

Rockwell International. *North Wind Eagle 3 Wind Turbine Generator.* Final Test Report. January 1980. RFP-3071/3533/80/9.

Rockwell International. *Operation of Small Wind Turbines on a Distribution System.* Two Vols. Executive Summary, March 1981, RFP-3177-1. Final Report, March 1981, RFP-3177-2.

Rockwell International. *Pinson C2E Wind Turbine Generator Failure Analysis and Corrective Design Modification.* March 1980. RFP-3148/3533/80-16.

Rockwell International. *Sencenbaugh-Model 1000-14 Wind Turbine Generator.* Performance Report. July 1979. RFP-3034/3533/79-5.

Rockwell International. *Small Wind Systems Technology Assessment. State of the Art and Near Term Goals.* February 1980. RFP-3136/3533/80/18.

Rockwell International. *Study of Dispersed Small Wind Systems Interconnected with a Utility Distribution System. Interim Report. Preliminary Hardware Assessment.* March 1980. RFP-3093/94445/3533/80/7. Contract #: DE-AC04-76DP03533. Prepared by Systems Control, Inc. for Rockwell International. Subcontract #: PF-94475.

Rockwell International. *The Technologies of Small Wind Energy Conversion Systems.* Technical Memorandum. March 1981. TM-TD/81-6.

Rockwell International. *Wind Machine Fatigue Analysis and Life Prediction.* April 1980. RFP-3135/3533/80-19.

Rockwell International. *Yaw Control Abstracts*. January 1980. Contract #: DE-AC04-76DP03533. Subcontract #: PF-01920T.

Rockwell International. *Zephyr Wind Dynamo Wind Turbine Generator*. Final Test Report. October 1979. RFP-3041/3533/79/13.

Technology Development—Large Machines

General Electric. *Design Study of Wind Turbines 50 kW to 3000 kW for Electric Utility Applications, Analysis and Design*. February 1976. ERDA/NASA-19403-76/2.

General Electric. *Executive Summary Mod-1 Wind Turbine Generator Analysis and Design Report*. March 1979. DOE/NASA/0058-79/3. NASA CR-159497.

General Electric. *System Dynamics of Multi-Unit Wind Energy Conversion Systems Application*. February 1978. DSE-2332-T2. Executive Summary DSE-2332-T1.

Gougeon Brothers, Inc. *Design and Proposed Fabrication for Low Cost Wood Composite Wind Turbine Blades*. Draft. July 1980. Contract (with NASA-Lewis Research Center): DEW3-101.

Hamilton Standard. *Experimental and Analytical Research on the Aerodynamics of Wind Turbines*. Final Report. December 1977. COO-2615-76/2.

Kaman. *Design, Fabrication, Test, and Evaluation of a Prototype 150-foot Long Composite Wind Turbine Blade*. September 1979. DOE/NASA/0600-79/1. NASA CR-159775.

Massachusetts Institute of Technology. *Wind Energy Conversion*. Vol. 1. Methods for Design Analysis of Horizontal Axis Wind Turbines. COO-4131-T1 (Vol. 1); Vol. 2. Aerodynamics of Horizontal Axis Wind Turbines. COO-4131-T2 (Vol. 2); Vol. 3. Dynamics of Horizontal Axis Wind Turbines. COO-4131-T1 (Vol. 3); Vol. 4. Drive System Dynamics. COO-4131-T2 (Vol. 4); Vol. 5. Experimental Investigation of a Horizontal Axis Wind Turbine. COO-4131-T1 (Vol. 5); Vol. 6. Nonlinear Response of Wind Turbine Rotor. COO-4131-T1 (Vol. 6); Vol. 7. Effect of Tower Motion on the Dynamic Response of Windmill Rotor. COO-4131-T1 (Vol. 7); Vol. 8. Free Wake Analysis of Wind Turbine Aerodynamics. COO-4131-T1 (Vol. 8); Vol. 9. Aerodynamics of Wind Turbine with Tower Disturbances. COO-4131-T1 (Vol. 9); Vol. 10. Aerolastic Stability of Wind Turbine Rotor Blades. COO-4131-T2 (Vol. 10). September 1978. Contract #: EY-76-S-02-1131.

NASA. *Comparison of Computer Codes for Calculating Dynamic Loads in Wind Turbines*. September 1977. DOE/NASA/1028-78/16. NASA TM-73773.

NASA. *Comparison of Upwind and Downwind Rotor Operations of the DOE/NASA 100-kW Mod-O Wind Turbine*. Prepared for Second DOE/NASA Wind Turbine Dynamics Workshop, Cleveland, Ohio, February 24-26, 1981. DOE/NASA/1028-31.

NASA. *Design, Fabrication and Initial Test of a Fixture for Reducing the Natural Frequency of the Mod-O Wind Turbine Tower*. July 1979. DOE/NASA/1028-79/24. NASA TM-79200.

NASA. *Design and Operating Experience on the U.S. Department of Energy Experimental Mod-0 100 kW Wind Turbine*. Paper presented at Thirteenth Intersociety Energy Conversion Engineering Conference, San Diego, California, August 20-25, 1978. DOE/NASA/1028-78/18. NASA TM-78915.

NASA. *Design Study of Wind Turbines 50 kW to 3000 kW for Electric Utility Applications*. Executive Summary and Vol. 2. July 1977. DOE/NASA/9404-76/1,2. NASA CR-134936.

NASA. *ERDA/NASA 100 kW Mod-0 Wind Turbine Operations and Performance*. Paper presented at the Conference on Wind Energy Conversion Systems, Washington, D.C., September 19-21, 1977. ERDA/NASA/1028-77/9. NASA TM-73825.

NASA. *Evaluation of Feasibility of Prestressed Concrete for Use in Wind Turbine Blades*. September 1979. DOE/NASA/5906-79/1. NASA CR-159725.

NASA. *Evaluation of Mostas Computer Code for Predicting Dynamic Loads in Two-Bladed Wind Turbines*. April 1979. DOE/NASA/1028-79/2. NASA TM-79101.

NASA. *Evaluation of Urethane for Feasibility of Use in Wind Turbine Blade Design.* April 1979. DOE/NASA/7653-79/1. NASA CR-159530.

NASA. *Fabrication and Assembly of the ERDA/NASA 100 kW Experimental Wind Turbine.* April 1976. DOE/NASA/1004-77/5.

NASA. *Feasibility Study of Aileron and Spoiler Control Systems for Large Horizontal Axis Wind Turbines.* May 1980. Wichita State University. DOE/NASA/3277-1.

NASA. *Investigation of Excitation Control for Wind-Turbine Generator Stability.* August 1977. ERDA/NASA/1023-77/3. August 1977. ERDA/NASA/1028-77/3. NASA TM-73745.

NASA. *Large Horizontal Axis Wind Turbine Development.* Prepared for Wind Energy Innovative Systems Conference sponsored by the Solar Energy Research Institute, Colorado Springs, Colorado, May 23-25, 1979. DOE/NASA/1059-79/2. NASA TM-79174.

NASA. *Large Wind Turbine Generators.* February 1978. DOE/NASA/1059-78/1. NASA TM-73767.

NASA. *Microprocessor Control of a Wind Turbine Generator.* Paper presented in Philadelphia, PA March 20-22, 1978. DOE/NASA/1028-78/20. NASA TM-79021.

NASA. *Mod-1 Wind Turbine Generator Analysis and Design Report.* March 1979. DOE/NASA/0058-79/2 or NASA CR-159495.

NASA. *Mod-1 Wind Turbine Generator Failure Modes and Effects Analysis.* February 1979. DOE/NASA/0058-79/1. NASA CR-159494,

NASA. *Mod-2 Failure Mode and Effects Analysis.* July 1979. DOE/NASA/0002-79/1. NASA CR 159632.

NASA. *Mod-2 Wind Turbine Farm Stability Study.* June 1980. NASA CR-16515. Contract #: DEN 3-134.

NASA. *Mod-2 Wind Turbine System Concept and Preliminary Design Report.* Executive Summary. July 1979. DOE/NASA/0002-80/2 or NASA CR-159609. Contract #: DEN 3-2.

NASA. *Mod-2 Wind Turbine System Concept and Preliminary Design Report.* Vol. 1. Executive Summary. Vol. II, Detailed Report. July 1979. DOE/NASA 0002-80/2. NASA CR-159609.

NASA. *Modified Aerospace Reliability and Quality Assurance Method for Wind Turbines.* Prepared for Annual Reliability and Maintainability Symposium, San Francisco, January 22-24, 1980. DOE/NASA/20370-79/18. NASA TM-79284.

NASA. *Modified Power Low Equations for Vertical Wind Profiles.* Prepared for Wind Characteristics and Wind Energy Siting Conference, Portland, Oregon, June 19-21. 1979. DOE/NASA/1059-79/4.

NASA. *NASTRAN Use for Cyclic Response and Fatigue Analysis of Wind Turbine Towers.* Reprinted from "Sixth NASTRAN User's Colloquium," October 4-6, 1977. ERDA/NASA/1004-77/3.

NASA. *Nonlinear Equations of Equilibrium for Elastic Helicopter or Wind Turbine Blades Undergoing Moderate Information.* December 1978. DOE/NASA/3082-78/1. NASA CR-159478.

NASA. *A 100 kW Experimental Wind Turbine: Simulation of Starting, Overspeed, and Shutdown Characteristics.* February 1976. DOE/NASA/1028-77/6.

NASA. *An Operating 200 kW Horizontal Axis Wind Turbine.* May 1978. DOE/NASA/1004-78/14. NASA TM-79034.

NASA. *Power Train Analysis for the DOE/NASA 100 kW Turbine Generator.* October 1978. DOE/NASA/1028-78/19. NASA TM-78997.

NASA. *Safety Considerations in the Design and Operation of Large Wind Turbines.* June 1979. DOE/NASA/20305-79/3. NASA TM-79193.

NASA. *Some Techniques for Reducing the Tower Shadow of the DOE/NASA Mod-0 Wind Turbine Tower.* September 1979. DOE/NASA/20370-79/17. NASA TM-79202.

NASA. *Survey of Long-Term Durability of Fiberglass-Reinforced Plastic Structures.* January 1981. DOE/NASA/9549-1. Prepared by Technical Report Services under Purchase Order C-39549-D.

NASA. *Teetered, Tip-Controlled Rotor: Preliminary Test Results from Mod-0 100-kW Experimental Wind Turbine.* Prepared for Wind Energy Conference, Boulder, Colorado, April 9–11, 1980. DOE/NASA/1028-80/26. NASA TM-81445.

NASA. *Test Evaluation of a Laminated Wood Wind Turbine Blade Concept.* May 1981. DOE/NASA/20320-30. Interagency Agreement #: DE-A101-76ET20320.

NASA. *Tower and Rotor Blade Vibration Test Results for a 100 kW Wind Turbine.* October 1976. DOE/NASA/1028-77/4.

NASA. *200 kW Wind Turbine Generator Conceptual Design Study.* January 1979. DOE/NASA/1028-79/1. NASA TM-79032.

NASA. *Utility Operational Experience on the NASA/DOE Mod-0A 200 kW Wind Turbine.* February 1979. DOE/NASA/1004-79/1. NASA TM-79084.

NASA. *Vibration Characteristics of a Large Wind Turbine Tower on Non-Rigid Foundations.* May 1977. ERDA/NASA 1004/77/1. NASA TM X-73670.

NASA. *Wake Characteristics of a Tower for the DOE-NASA Mod-1 Wind Turbine.* April 1978. DOE/NASA/1028-78/17. NASA TM-78853.

NASA. *Wind Turbine for Electric Utilities: Development Status and Economics.* DOE/NASA/1028-79/23. NASA TM-79170.

NASA. *Wind Turbine Generator Rotor Blade Concepts with Low Cost Potential.* May 1978. DOE/NASA/1028-77/13. NASA TM-73835.

Paragon Pacific Incorporated. *Coupled Dynamics Analysis of Wind Energy Systems.* January 1977. DOE/NASA/9767-77/1.

Paragon Pacific. *Wind Energy System Time-Domain (West) Analyzers Using Hybrid Simulation Techniques.* October 1979. DOE/NASA/0026-79/1. NASA CR-159737.

Structural Composites Industries, Inc. *Design and Evaluation of Low Cost Composite Blades for Large Wind-Driven Generating Systems. Monthly Technical Progress Narrative Report #18.* September 1980. Prepared for NASA-Lewis Research Center. Contract #: DEN3-100.

United Technologies. *Design of a Self-Regulating Composite Bearingless Blade Wind Turbine.* Final Report for October 15, 1976–August 15, 1977. January 1978. COO-4150-77/8.

Vertical Axis Wind Turbines

Alcoa Laboratories. *Design and Fabrication of a Low Cost Darrieus Vertical-Axis Wind Turbine System. Phase I—Technical Report.* March 1980. ALO-4272. Contract #: EM-78-C-04-4272.

Sandia. *Accelerometer Measurements of Aerodynamic Torque on the DOE/Sandia 17-M Vertical Axis Wind Turbine.* April 1981. SAND80-2776.

Sandia. *Aerodynamic Characteristics of Seven Symmetrical Airfoil Sections Through 180-Degree Angle of Attack for Use in Aerodynamic Analysis of Vertical Axis Wind Turbines.* March 1981. SAND80-2114.

Sandia. *Aerodynamic Interference Between Two Darrieus Wind Turbines.* April 1981. SAND81-0896.

Sandia. *Aerodynamic Performance of a 5-Metre Diameter Darrieus Turbine with Extruded Aluminum NACA-0015 Blades.* March 1980. SAND80-0179.

Sandia. *Aerodynamic Performance of the 17-Metre-Diameter Darrieus Wind Turbine.* January 1979. SAND78-1738.

Sandia. *Aerodynamic Performance of the 17-M-Diameter Darrieus Wind Turbine in the Three-Bladed Confuguration: An Addendum.* February 1980. SAND79-1753.

Sandia. *Aerodynamic Sizing of Vertical Axis Wind Turbines for Wind Farms.* August 1981. SAND81-0979.

Sandia. *Aerolastic Analysis of the Troposkein-Type Wind Turbine.* April 1977. SAND77-0026.

Sandia. *Application of the Darrieus Vertical-Axis Wind Turbine to Synchronous Electrical Power Generation.* March 1975. SAND75-0165.

Sandia. *An Automatic-Control System for the 17-Metre Vertical-Axis Wind Turbine (VAWT).* March 1980. SAND78-0984.

Sandia. *Comparison with Strain Gage Data of Centrifugal Stresses Predicted by Finite Element Analysis on the DOE/Sandia 17-M Darrieus Turbine.* February 1980. SAND79-1990.

Sandia. *Characteristics of Future Vertical-Axis Wind Turbines.* July 1978. SAND79-1068.

Sandia. *The Darrieus Turbine: A Performance Prediction Model Using Multiple Streamtubes.* October 1975. SAND75-0431.

Sandia. *Economic Analysis of Darrieus Vertical-Axis Wind Turbine Systems for the Generation of Utility Grid Electrical Power.* Vols. I-IV. August 1979. SAND78-0962.

Sandia. *Engineering Development Status of the Darrieus Wind Turbine.* March 1977. SAND76-0650.

Sandia. *Fixed Wing Analysis of the Darrieus Rotor.* July 1981. SAND81-7026.

Sandia. *Fourier Coefficients of Aerodynamic Torque Functions for the DOE/Sandia 17-M Vertical-Wind Turbine.* February 1980. SAND79-1508.

Sandia. *FY 79 Program Plan Technical and Management Support for the Vertical-Axis Wind Turbine Program.* November 1979. SAND79-1594.

Sandia. *Guy Cable Design and Dumping for Vertical Axis Wind Turbines.* May 1981. SAND80-2669.

Sandia. *Induction and Synchronous Machines for Vertical Axis Wind Turbines.* June 1979. SAND79-7017.

NASA. *Nonlinear Aerolastic Equations of Motion of Twisted Nonuniform Flexible Horizontal-Axis Wind Turbine Blades.* July 1980. University of Toledo. DOE/NASA/3139-1.

Sandia. *Nonlinear Stress Analysis of Vertical-Axis Wind Turbine Blades.* April 1975. SAND74-0378.

Sandia. *Preliminary Blade Strain Gage on the Sandia 17-Meter Vertical-Axis Wind Turbine.* December 1977. SAND77-1176.

Sandia. *Proceedings of the Vertical Axis Wind Turbine (VAWT) Design Technology Seminar for Industry.* April 1-3, 1980, Albuquerque, New Mexico. August 1980. SAND80-098.

Sandia. *Residual Stresses in Darrieus Vertical Axis Wind Turbine Blades.* April 1981. SAND81-0923.

Sandia. *A Study of Foundation/Anchor Requirements for Prototype Vertical-Axis Wind Turbines.* February 1979. SAND78-7046.

Sandia. *Torque Ripple in a Darrieus Vertical Axis Wind Turbine.* September 1980. SAND80-0475.

Sandia. *A User's Manual for the Vertical Axis Wind Turbine Code VDART 3.* June 1981. SAND81-7020.

Sandia. *A Vortex Model of the Darrieus Turbine: An Analytical and Experimental Study.* June 1981. SAND81-7017.

Tetra Tech, Inc. *Augmented Horizontal Axis Wind Energy Systems Assessment.* December 1979. Executive Summary, Final Report. SERI/TR-98003-3.

Toledo, University of. *Aerolastic Equations of Motion of a Darrieus Vertical-Axis Wind Turbine Blade.* December 1979. DOE/NASA/1028-79/25. NASA TM-79295.

West Virginia University. *The Effects of Flow Curvature on the Aerodynamics of Darrieus Wind Turbines.* July 1980. ORO-5135-77/7.

West Virginia University. *Vertical-Axis Wind Turbine Development.* July 1979. ORO-5135-77/5.

Innovative Systems

AeroVironment, Inc. *Advanced and Innovative Wind Energy Concept Development: Dynamic Inducer System.* Research Report. May 1981. SERI/TR-8085-1-T2.

AeroVironment, Inc. *A Definitive Generic Study of Augmented Horizontal Axis Wind Energy Systems*. April 1979. SERI/TR-98003-1.

Dayton, University of. *An Analysis of the Maderas Rotor Power Plant—An Alternate Method for Extracting Large Amounts of Power from the Wind*. Vol. 2. Technical Report. June 1978. DSE-2554-78/2.

Dayton, University of. *Third Annual Progress Report on the Electrofluid Dynamic Wind Generator*. Final Report. September 15, 1977–September 30, 1978. May 1979. COO-4130-2. (See SERI in this section for listing of *Fourth Annual Progress Report*.)

Grumman. *Investigation of Diffuser-Augmented Wind Turbines*. Parts I & II. January 1977. ERDA COO-2616-2.

Grumman. *Investigation of the Tornado Wind Energy Systems*. Draft. Final Report for period of September 1978 through April 1980. May 1980. DOE Contract #: E(49-18)2555.

Grumman. *A Non-Aerospace Application of Plans: Preliminary Structural Design of Wind Turbine Diffuser*. March 1977. RM-629.

Grumman. *Subsonic Diffuser with Injection Flow Along the Boundary*. October 1977. RM-644.

JBF. *Preliminary Technical and Economic Evaluation of Vortex Extraction Devices*. April 1980. Summary Report. SERI/TR-8003-2; Final Report. SERI/TR-8003-1.

McDonnell Aircraft Co. *Feasibility Investigation of the Giromill for Generation of Electrical Power*. Volume 1 (Executive Summary) and Volume 2 (Technical Discussion). December 1977. COO/2617/76/1/2.

Montana State University. *Technical Feasibility Study of a Wind Energy Conversion System Based on the Tracked Vehicle-Airfoil Concept*. September 30, 1974. NSF/RANN/SE/GI-39415/PR/74/3.

Polytechnic Institute of New York. *Vortex Augmentors for Wind Energy Conversion Progress Report, May-November 1976*. December 1976. TID-27885.

Princeton University. *Optimization and Characteristics of a Sailwing Windmill Rotor*. July 31, 1974. NSF/RANN/SE/GI-41891/PR/74/2.

Sandia. *Wind Tunnel Performance Data for Two- and Three-Bucket Savonius Rotors*. July 1977. SAND76-0131.

SERI. *A Definitive Generic Study for Sailwing Wind Energy Systems*. Non-Technical Summary Report. Final Report. October 1979. SERI/TR98003-05.

SERI. *Fourth Annual Progress Report on the Electrofluid Dynamic Wind Generator*. Final Report for the Period 1 April 1979-31 August 1980. (Under sub-contract with University of Dayton)

SERI. *Giromill Overview*. May 23-25, 1979. SERI/TP-35-263.

SERI. *A Review of the Current Status of the Wind Energy Innovative Systems Projects*. March 1980. SERI/TP-635-469.

Wind Characteristics

Battelle. *Annual Report of the Wind Characteristics Program Element for the Period July 1978 through September 1979*. May 1980. PNL-3211.

Battelle. *A Practical Method for Estimating Wind Characteristics at Potential Wind Energy Conversion Site*. PNL-3808. Prepared for Battelle by SRI International under Agreement #B-23149-A-L.

Battelle. *Accuracy of Wind Power Estimates*. October 1, 1977. Contract #EY-76-C-06-1830. PNL-2442.

Battelle. *Assessing the Local Wind Field with Instrumentation*. October 1980. PNL-3622. Prepared for Battelle by AeroVironment, Inc. under Agreement B-92864-A-H.

Battelle. *Candidate Wind Turbine Generator Site Annual Data Summary for January 1980 Through December 1980*. April 1981. PNL-3739.

Battelle. *Estimates of the Number of Large Amplitude Gusts*. March 1978. PNL-2508.

Battelle. *Estimation of Wind Characteristics at Potential Wind Energy Conversion Sites*. October 1979. PNL-3074.

Battelle. *A Measurement Program to Characterize the Wind at a Potential WECS Site.* March 1978. PNL-2516.

Battelle. *Preliminary Evaluation of Wind Energy Potential-Cook Inlet Area, Alaska,* June 1980. Contract #: DE-AC06--76 RLO 1830.

Battelle. *Report from a Working Group Meeting on Wind Forecasts for WECS Operation.* March 1978. Contract #EY-76-C-06-1830.

Battelle. *Simulation of the Hourly Wind Speeds for Randomly Dispersed Sites.* May 1978. Contract #EY-76-C-06-1830. PNL-2523.

Battelle-PNL. *A Siting Handbook for Small Wind Energy Conversion Systems.* March 1980. Contract #: DE-AC06-76 RLO 1830. PNL-2521. Rev. 1.

Battelle. *Some Aspects of Fluctuating Vertical Wind Shears.* May 1981. PNL-3771.

Battelle. *Summary of Wind Data from Nuclear Power Plant Sites.* March 1977. BNWL-WIND-4.

Battelle. *Survey of Historical and Current Site Selection Techniques for the Placement of Small Wind Energy Conversion Systems.* December 1977. BNWL-2220 WIND-9.

Battelle. *Synthesis of National Wind Energy Assessments.* July 1977. BNWL-2220 WIND-5.

Battelle. *Wind Direction Change Criteria for Wind Turbine Design.* January 1979. PNL-2531.

Battelle. *Wind Energy Resource Altases;* 12 Vols. Contract # for all volumes: DE-AC06-76RLO 1830. *Vol. 1—The Northwest Region.* April 1980. PNL-315 WERA-1. *Vol. 2—The North Central Region.* February 1981. PNL-3195 WERA-2. Prepared for PNL by ERT/Western Scientific Services, Inc. under Agreement B-87926-A-1; *Vol. 3—The Great Lakes Region.* February 1981. PNL-3195 WERA-3. Prepared for PNL by Environmental Research and Technology, Inc. under Agreement B-87922-A-L. *Vol. 4—The Northeast Region.* September 1980. PNL-3195 WERA-4. Prepared for PNL by GEOMET Technologies, Inc., under Agreement B-23452-A-L; *Vol. 5—The East Central Region.* PNL-3195 WERA-5. Prepared for PNL by NUS Corporation under Agreement B-87924-A-L; *Vol. 6—The Southeast Region.* January 1981. PNL-3195 WERA-6. Prepared for PNL by GEOMET Technologies, Inc., under Agreement B-87925-A-L; *Vol. 7—The South Central Region.* March 1981. PNL-3195 WERA-7. Prepared for PNL by Institute for Storm Research under Agreement B-87923-A-L; *Vol. 8—The Southern Rocky Mountain Region.* March 1981. PNL-3195 WERA-8. Prepared for PNL by ERT/Western Scientific Services, Inc., under Agreement B-87921-A-L; *Vol. 9–The Southwest Region.* November 1980. PNL-3195 WERA-9. Prepared for PNL by Global Weather Consultants, Inc., under Agreement B-87920-A-L; *Vol. 10—The Alaska Region.* December 1980. PNL-3195 WERA-10. Prepared for PNL by University of Alaska under Agreement B-87917-A-L. *Vol. 11–Hawaii and Pacific Islands Region.* February 1981. PNL 3195 WERA 11. Prepared for PNL by University of Hawaii under Agreement B87918-A-L. *Vol. 12—Puerto Rico and U.S. Virgin Islands.* January 1981. PNL-3195 WERA-12. Prepared under Contract #DE-AC06-76RLO 1830.

Battelle. *Wind Measurement Systems and Wind Tunnel Evaluation of Selected Instruments.* May 1981. PNL-3435.

Battelle. *Wind Velocity-Change (Gust Rise) Criteria for Wind Turbine Design.* Contract #EY-76-C-06-1830. PNL-2526.

Colorado State University. *Sites for Windpower Installations: Physical Modeling of the Influence of Hills, Ridges and Complex Terrain on Wind Speed and Turbulence.* June 1978. RLO-2438-78/1 (Executive Summary); RLO-2438-73/2 (Final Report); RLO-2438-78/3 (Appendices).

FWG. *Summary of Guidelines for Siting Wind Turbine Generators Relative to Small-Scale, Two-Dimensional Terrain Features.* March 1979. RLO/2443-77/1.

Northrop Services, Inc. *Site Insolation and Wind Power.* Technical Report Southern Region (Vol. 3); Technical Report Midwest Region (Vol. 4); Technical Report Western Region (North Section) (Vol. 5); Technical Report Western Region (South Section) (Vol. 6). Published August 1980. DOE/CS/20160-01.

Northwestern University. *Application of Statistical Techniques to Wind Characteristics at Potential Wind Energy Conversion Sites.* Final Report for the Period October 1, 1978–September 30, 1979. May 1980. DOE/ET/20283-2.

Northwestern University. *Statistical Models for Wind Characteristics at Potential Wind Energy Conversion Sites.* January 1979. DOE/ET/20203-1.

Northwestern University. *Stochastic Modelling of Site Wind Characteristics.* September 1977. RLO/2342-77/2.

Oregon State University. *Analysis of Strong Nocturnal Shears for Wind Machine Design.* Progress Report. December 1979. DOE/ET-23116-79-1.

Oregon State University. *A Handbook on the Use of Trees as an Indicator of Wind Power Potential.* June 1979. RLO-2227-79/3.

Oregon State University. *Vegetation as an Indicator of High Wind Velocity.* June 1978. RLO-2227-T24-78-2.

Poseidon Research. *The Effect of Atmospheric Density Stratification on Wind Turbine Siting.* Final Report. January 1978. RLO-2444-78/1.

Research Triangle Inst. *On a Technique to Determine Wind Statistics in Remote Locations.* Final Report. December 1977. RLO-2445-78/1.

Sandia. *Some Variability Statistics of Available Wind Power.* March 1979. SAND78-1735.

Sandia. *Wind Characteristics for Field Testing of Wind Energy Conversion Systems.* November 1979. SAND78-1563.

SERI. *County-Level Wind Resource Estimates.* Final Report. February 1981. SERI/TR-9095-1. Prepared by PRC under Subcontract #: XE-0-9095-1.

SERI. *Near-Term High Potential Counties for SWECS. Final Report.* February 1981. SERI/TR-98282-11. Contract #: EG-77-C-01-4042. (Prepared by A. D. Little, Inc. under Subcontract #BE-9-8282-11).

South Dakota School of Mines. *Energy From Humid Air.* August 1977. DSE/2553-77/1.

Tennessee, University of. *Engineering Handbook on the Atmospheric Environmental Guidelines for Use in Wind Turbine Generator Development.* July 1977. University of Tennessee Space Institute. NAS8-32118.

Virginia Polytechnic Institute and State University. *Planetary Boundary Layer Wind Model Evaluation at a Mid-Atlantic Coastal Site.* October 1980. DOE/ET/23007-80/1.

Virginia, University of. *Coastal Wind Energy.* Pt. 1: Synthesis and Results; Pt. 2: Climatology. January 1978. RLO-2344-76/77-5.

Wyoming, University of. *A Guide of the Interpretation of Windflow Characteristics from Eolian Landforms.* April 1979. RLO-2343-79/2.

Legal, Social, Environmental Issues

Booz-Allen. *Economic Incentives to Wind Systems Commercialization.* Final Report and Executive Summary. August 1978. DOE/ET/4053-78/1.

DOE. *Environmental Development Plan, Wind Energy Conversion.* 1977. DOE/EDP-0007.

DOE. *Final Environmental Impact Statement: Wind Turbine Generator System, Block Island, Rhode Island.* July 1978. DOE/EIS-0006.

DOE. *A Preliminary Analysis of the Audible Noise of Constant-Speed, Horizontal-Axis Wind-Turbine Generators.* July 1980. DOE/EV-0089.

ERCO. *A Market Analysis of the Potential for Wind Systems Use in Remote and Isolated Area Applications.* June 8, 1979. HQS/4051-77/9.

Flow Research Company. *A Review of Wind Turbine Wake Effects.* January 1980. RLO/3160-80/1.

General Motors. Small *Wind Turbine Production Evaluation and Cost Analysis, Interim Report Wind Turbine Design Analysis.* November 1980. EP80 123. SERI/RR-9049-1.

George Washington University. *Legal-Institutional Arrangements Facilitating Off-shore Wind Energy Conversion Systems (WECS) Utilization.* Final Report. September 1977. DOE/NSF/19137-77/3.

JBF. *Summary of Current Cost Estimates of Large Wind Energy Systems.* February 1977. DSE/2521-1.

Michigan State University. *Impact of Storm Fronts on Utilities with WECS Arrays.* October 1979. COO/4450-79/2.

Michigan, University of. *Electromagnetic Interference by Wind Turbine Generators.* March 1978. TID-28828.

Michigan, University of. *Wind Turbine Generator Siting and TV Reception Handbook.* Technical Report No. 1. January 1978. COO-2846-1.

NASA. *Lightning Accommodation Systems for Wind Turbine Generator Safety.* Prepared for Fifth International System Safety Conference, Denver, Colorado, July 26–31, 1981. DOE/NASA/20320-31.

Nevada, University of. *Icing on Wind Energy Systems.* January 1981. DOE/ET/23170-89. Contract #: AC06-78ET23170.

Rockwell International. *Current State Incentive Programs for Small Wind Energy Conversion Systems.* Final Report. July 1980. RFP-3128/05480/3533/80-0.

Rockwell International. *Financial Problems Facing the Manufacturers of Small Wind Energy Conversion Systems.* November 1979. DOE/DP-03533-T2.

Rockwell International. *Issues and Examples of Developing Utility Interconnection Guidelines for Small Power Production.* Technical Memorandum. January 1981. TM-IP/81-5.

Rockwell International. *Small Wind Energy Conversion Systems: Zoning Issues and Approaches.* May 1981. TM-IP/81-7.

Rockwell International. *SWECS Cost of Energy Based on Life Cycle Costing.* Technical Report. May 1980. RFP-3120/3533/80-13.

SERI. *Capital Formation for Small Wind Energy Conversion System Manufacturers, A Guide to Methods & Sources.* Final Report. May 1980. SERI/TR-98298-1. Subcontract #: AM-9-8298-1.

SERI. *Economics of Selected WECS Dispersed Applications.* April 1980. SERI/TR-431-580.

SERI. *Electric Utility Value Determination for Wind Energy. Volume II: A User's Guide.* February 1981. SERI/TR-732-604. Contract #: EG-77-C-01-4042.

SERI. *A General Reliability and Safety Methodology and Its Application to Wind Energy Conversion Systems.* September 1979. SERI/TR-35-234.

SERI. *Land-Use Implications of Wind Energy Conversion Systems.* February 1981. SERI/TP-744-1099.

SERI. *Legal and Institutional Implications of Providing Financial Incentives to Encourage the Development of Solar Technologies.* July 1979. SERI/TR-62-269.

SERI. *Product Liability Insurance for WECS.* October 1979. SERI/TR-354-466.

SERI. *Product Liability and Small Wind Energy Conversion Systems (SWECS): An Analysis of Selected Issues and Policy Alternatives.* December 1979. SERI/TR-354-365.

SERI. *Solar Energy Legal Bibliography, Final Report.* March 1979. SERI/TR-62-069.

SERI. *Utility Rates and Service Policies as Potential Barriers to the Market Penetration of Decentralized Solar Technologies.* August 1979. SERI/TR-62-274.

SERI. *Utility Siting of WECS: A Preliminary Legal/Regulatory Assessment.* May 1981. SERI/TR-744-778.

COMMERCIAL WIND MACHINES

This list was prepared by the Rocky Flats Wind Systems Program as part of their contract to the U.S. Department of Energy. The principal author is Darrell Dodge. It is current as of 1 March 1981.

Because of the time involved in publication and the volatile nature of the wind industry, it is extremely likely that this list will be incomplete, not including new companies, and including companies who have since gone out of business, by the time of publication.

Commercially Available Wind Machines (Electrical)

Name and Model Number	Rated Output (kw @ mph)	Diameter (ft/m)
Alcoa Allied Products—ALVAWT 835524*	50 kW/30 mph	55/17
Kaman Aerospace—Kaman 40	40 kW/20 mph	64/19.5
Merkam Energy Development Co.—440	40 kW/25 mph	40/12.2
McDonald Aircraft Co.—Giromill*	40 kW/20 mph	58/17.7
Lebost Turbines, Inc.—LWT-26	26 kW/26 mph	26/7.7
Jay Carter—Mod 25	25 kW/25 mph	32/9.8
Wind Engineering—Windgen 25	25 kW/25 mph	38/11.6
Fayette Manufacturing—Winway	20 kW/50 mph	13.5/4.3
Tumac Industries—10-meter*	20 kW/n.a.	32.8/10
Wind Power Systems, Inc.—Storm Master	18 kW/24 mph	32.8/10
Coulson Wind Electric—Zephyr	15 kW/25 mph	21/6.4
Environmental Energies—HWT 15	15 kW/25 mph	24/7.3
Environmental Energies—HWT 12	12 kW/25 mph	22/6.7
Lebost Turbines, Inc.—LWT 21	12 kW/26 mph	21/6.4
Windworks, Inc.—Windworker-10	10 kW/20 mph	32.5/10
Astral-Wilcon—AW 10-B	10 kW/22 mph	26/7.7
Jacobs Wind Electric Co.	10 kW/25 mph	23/7
Millville Windmills—10-3-IND	10 kW/25	25/7.6
Elektro—WVG 120G	10 kW/31	21.7/6.6
Environmental Energies—HWT 9	9 kW/25 mph	20/5.1
Wind Power Systems—10-9-1G-1P-60	9 kW/20 mph	32.8/10
Lebost Turbines, Inc.—LWT-14	8 kW/26 mph	14/4.3
Wind Power Systems—10-8-BC-48-PM	8 kW/20 mph	32.8/10
Tumac Industries—Model 7050*	6.8 kW/17.5 mph	16.4/5
Environmental Energies—HWT 6	6 kW/25 mph	18/5.5
Delta—T—Aero Polyblade	6. kW/34 mph	22/6.6
Elektro—WVG 50G	5.4 kW/27 mph	16.4/5
TWR Enterprises—Wind Titan	5 kW/25 mph	18/5.5

(*Continued*)

* denotes vertical axis

Commercially Available Wind Machines (Electrical) *(Cont.)*

Name and Model Number	Rated Output (kw @ mph)	Diameter (ft/m)
Kedco, Inc.—1840	5 kW/25 mph	18/5.5
Tumac Industries—5-meter*	5 kW/n.a.	16.4/5
American Energy Savers—Reinke	5 kW/24 mph	19/5.8
Dunlite Electrical Products	5 kW/34 mph	18.2/5.6
Enertech Corp.—4000	4 kW/n.a.	19.7/6
Independent Energy Systems—Skyhawk	4 kW/23 mph	15/4.6
The Gale Company	4 kW/14 mph	39/12
Aerowatt S.A.—4100FP&G	4.1 kW/16 mph	32.7/10
Hummingbird Windpower Corp.—4000/22	4 kW/22 mph	14/4.3
Whirlwind Power systems—Model AA	4 kW/25 mph	18/5.5
Elektro—WV 35G	3.8 kW/27 mph	14.5/4.4
Dynergy—5-meter*	3.3 kW/24 mph	15/4.6
Hinton Research—3A, B, C	3 kW/29 mph	11/3.4
Product Development Institute	3 kW/27 mph	13.6/4.15
TWR Enterprises—Wind Titan	3 kW/25 mph	10/3.1
Product Development Institute—Wind Jennie	3 kW/25 mph	12.5/3.8
Kedco, Inc.—1620	3 kW/25 mph	16/4.88
Birch Machine, Inc.—Windcraft	2.5 kW/n.a.	15/4.6
Lebost Turbines, Inc—LWT-8	2.5 kW/26 mph	8/2.4
Altos—BWP-12A	2.2 kW/38 mph	8/2.4
North Wind Power Co.—HR2	2 kW/20 mph	16.4/5
Pinson Energy Corp.—C-2E*	2 kW/24 mph	12/3.6
Altos—BWP-12B	2 kW/28 mph	11.5/3.5
Bertoia Studio—A.P.S.	2 kW/18 mph	18/5.49
Dunlite—81/002550	2 kW/25 mph	13.5/4.1
Kedco, Inc.—1210	2 kW/26 mph	12/3.65
Kedco, Inc.—1610	2 kW/22 mph	16/4.88
Sencenbaugh—2000/115	2 kW/26 mph	12/3.65
Whirlwind Power Co.—Model A	2 kW/25 mph	12/3.65
Windflo—Wingen Mod 2000	2 kW/25 mph	10/3
Enertech Corp.—1800	2 kW/24 mph	13/4
Kedco, Inc.—1605	1.9 kW/20 mph	16/4.88
Enertech Corp.—1500	1.5 kW/21 mph	13.2/4
Altos—BWP-8A	1.5 kW/28 mph	8/2.4
Tumac Industries—5 meter	1.5 kW/n.a.	16.4/5
Elektro—WV 15G	1.2 kW/27 mph	10/3
Kedco, Ind.—1200	1.2 kW/22 mph	12/3.65
TWR Enterprises—Wind Titan	1.2 kW/25 mph	14/4.3
Aerowatt S.A.—1100PF7G	1.1 kW/16 mph	16.7/5
Aero Power systems, Inc.—SL1000	1 kW/20 mph	10/3.05
Dergey Wind Power—BWE 1000	1 kW/25 mph	8.5/2.6
Dunlite Electrical—81/002550	1 kW/37 mph	10/3.1
Chalk Wind Systems—360TLN	1 kW/30 mph	11.5/3.5
Sancken Wind electric	1 kW/30 mph	n.a.
Sencebaugh Wind Electric—1000	1 kW/23 mph	12/3.65
Bergey Wind Power—BWC 650	0.65/25 mph	8.5/2.6

*vertical axis

Electrical Wind Machine Manufacturers

Aero Power
2398 Fourth St.
Berkeley, CA 94710
(415) 848-2710

Aerowatt S.A.
c/o Automatic Power, Inc.
P.O. Box 18738
Houston, TX 77023
(713) 228-5208

Alcoa Allied Products
Alcoa Laboratory ATC-A
Alcoa Center, PA 15069
(412) 337-2977

Altos
P.O. Box 905
Boulder, CO 80302
(303) 442-0885

American Energy Savers
912 St. Paul Rd.
Box 1421
Grand Island, NE 68801
(308) 382-1831

Astral/Wilcon
P.O. Box 291
Milbury, MA 01527

Bergey Wind Power Co.
2001 Priestly Ave.
Norman, OK 73069
(405) 364-4212

Bertoia Studia
644 Main St.
Bally, PA 19503
(215) 845-7096

Jay Carter Enterprises
P.O. Box 684
Burkburnett, TX 76354
(817) 569-2238

Chalk Wind Systems
P.O. Box 446
St. Cloud, FL 32769
(305) 892-7338

Coulson Wind Electric Inc.
RFD 1, Box 225
Polk City, IA 50226
(515) 984-6038

Dakota Wind & Sun
P.O. Box 1781
Aberdeen, SD 57401
(605) 229-0815

Dragonfly Wind Electric
P.O. Box 155
Elk, CA 95432
(707) 877-3474

Dunlite Electrical Products
Enertech Corp.
P.O. Box 420
Norwich, VT 05055
(802) 649-1145

Dynergy Corp.
P.O. Box 428
Laconia, NH 03246
(603) 524-8313

Enertech Corp.
P.O. Box 420
Norwich, VT 05055
(802) 649-1145

Environmental Energies, Inc
Front St.
Copemish, MI 49625
(616) 378-2000

Fayetter Manufacturing Corp.
RD 1, Box 262C
Kersey, PA 15846
(814) 885-8569

The Gale Company
P.O. Box 27
Lake Geneva, WI 53147
(414) 248-6672

Grumman Energy Systems
4175 Veterans Memorial Hwy.
Ronkonkoma, NY 11779
(516) 575-6205

Hinton Research
417 Kensington
Salt Lake City, UT 84115
(801) 487-3869

Hummingbird Windpower Corp.
P.O. Box 1248
Sweetwater, TX 79556
(915) 235-1735

Independent Energy Systems
6043 Sterrettania Rd.
Fairview, PA 16415
(814) 833-3567

Jacobs Wind Electric
2180 W. 1st St. Suite 410
Fort Meyers, FL 33901
(813) 481-3113

Kaman Aerospace Corp.
Old Windsor Rd.
Bloomfield, CT 06002
(203) 242-4461

Kedco, Inc.
9016 Aviation Blvd.
Inglewood, CA 90301
(213) 776-6636

Lebost Turbines, Inc
1116 Wharburton Ave.
Yonkers, NY 10701
(914) 423-8414

McDonnell Aircraft Co.
P.O. Box 516
St. Louis, MO 63166
(314) 232-7998

Megatech Corp.
29 Cook St.
Billerica, MA 01866
(617) 273-1900

Merkham Energy Development Co.
179 East Rd. #2
Hamburg, PA 19526
(215) 562-8856

(Continued)

Electrical Wind Machine Manufacturers (*Cont.*)

Millville Windmills, Inc.
10335 Old Dr.
Millville, CA 96062
(916) 547-4302

North Wind Power Co.
Box 556
Moretown, VT 05660
(802) 496-2955

Pinson Energy Corp.
P.O. Box 7
Marston Mills, MA 02648
(617) 477-2913

Product Development Institute
508 S. Byrne Rd.
Toledo, OH 43609
(419) 382-0282

Sacken Wind Electric
4140 Skylark
Kingman, AZ 86401
(602) 757-2526

Sencenbauch Wind Electric
P.O. Box 11174
Palo Alto, CA 94306
(415) 964-1593

Tumac Industries
650 Ford St.
Colorado Springs, CO 80915
(303) 596-4400

TWR Enterprises
72 W. Meadow La.
Sandy, UT 84070

Windworks, Inc
Rt. 3 Box 44A
Mukwonago, WI 53149
(414) 363-4088

Whirlwind Power Co.
2458 W. 29th Ave.
Denver, CO 80211
(303) 477-6436

Wind Power Systems
8030 Production Ave.
San Diego, CA 92121
(714) 566-1806

Windflo Power Ltd.
90 Esna Dr.
Markham, Ontario
Canada L3R 2R7

INDEX

active power, 68
aesthetics, 76
air brake, xvii
air break disconnect, 66
alternating current, 41
alternator, synchronous, 71
American National Standards Institute, 47
American Planning Association, 75
American Society for Testing and Materials, 47
American Society of Mechanical Engineers, 47
American Wind Energy Association, xxi, xxii, 43, 47
appliances, household, 5
Asimov, Isaac, xxiii
average wind speed, 24, 32
avoided cost, 62-64
axis, vertical, horizontal, 41

Battelle Pacific Northwest Laboratories, 24
batteries, xv, xvii
bird safety, 80
biomass, 2, 61
Boone, North Carolina, 79
blades, material, number of, 41
   throw, 75, 77
building codes, 77-78
buy-back rate, 6, 34, 62-64

California, 10
cash flow, 56-59
   discounted, 54
cogenerate, 61
controlled velocity tests, 42
controls, overspeed, xvii, 35, 37-40
Cooke, Alistair, 60
cube relationship, 12
cumulative breakeven, 57
current, in-rush, 70
cut-in wind speed, 35, 40, 45
cut-out wind speed, 35, 40, 45

design competition, xix
design output, 35
direct current, 41
direct drive, xvii, 41
discount rate, 53

economic utilization factor, 56
Energy, Department of, xix

feathering, 39
Federal Energy Regulatory Commission, 61
fixed charge rate, 55
flagging, 27, 31-32
flyball hub, xviii
Fuller, R. Buckminster, xii

gearbox, 17
General Electric, 5
generator, 41
   induction, 66, 71

Hansen, A. C., 79
harmonics, 67-69
home-built, xiii
hydropower, 2, 61

ice, 77
insurance, 72
inverter, line-commutated, 61, 71
   self-commutated, 61, 71

kinetic energy, 12, 15

land use, 14
life cycle cost, 51
lightening, 67
load, negative, 66
   factor, 20

181

marginal cost, xxiii, 61
Meyer, Hans, xii
Mod-1, 79

National Climatic Center, 25
National Electrical Code, 48, 66
National Fire Protection Association, 47
neighbors, 74, 81
noise, 16, 79

obstructions, 10
ocean thermal energy conversion, xix

payback, 56-59
photovoltaics, xix, 61
power
   quality of, 67
   reactive, 68
power curve, 42, 44-45
power factor, 67
   lagging, 68
   leading, 68
Public Utility Regulatory Policies Act, 61-64
Putnam, Palmer, 27

quad, 3, 4, 5

rated output, 31, 37, 41
Rayleigh distribution, 45, 46
reliability, 37
remote applications, 91
Reuss, Henry, xix
ridges, 10, 13
Rocky Flats, 37, 41-43, 79
rotor diameter, 31, 40-41, 35
rural electric cooperatives, 60
Rural Electricifcation Administration, xii, 60

sea breeze, 8
sensivity analysis, 56
setbacks, 75-76

slip rings, xv
small power production, 61
solar, 1, 2, 3, 61
stall, 39
surface roughness, 10
surge arrestors, 67

tax credits, 59
television interference, 16, 78-79
tip flaps, 39
Total Harmonic Distortion, 69
Tower height, 74-75
   foundation, 85
trade winds, 17
transmission, xvii, 4, 41
tree deformation, 27
turbulence, 10, 11, 42, 45, 74-76

Uniform Building Code, 77

variable axis rotor, xiii
variance, 81
vars, 68, 70
Venturi, xii
velocity distribution, 45
vertical wind gradient, 10, 12
vibration, 86
voltage flicker, 70

weather patterns, 27
Wendelgass, Paul, 75
Wilkerson, Alan, xvii
wind
   power density, 13, 25
   rights, 76
wind farms, xxii, 10, 14, 16, 93
Wind Resource Atlas, 24-31
Windworks, Inc., xvi, xii

zoning boards, 1, 6, 74, 80-83